MOBILE URBANISM

GLOBALIZATION AND COMMUNITY

SUSAN E. CLARKE AND GARY GAILE, SERIES EDITORS
DENNIS R. JUDD, FOUNDING EDITOR

(continued on page 215)

GLOBALIZATION AND COMMUNITY, VOLUME 17

Mobile Urbanism

CITIES AND POLICYMAKING IN THE GLOBAL AGE

Eugene McCann and Kevin Ward, Editors

Foreword by Allan Cochrane

University of Minnesota Press
Minneapolis
London

MINNESOTA

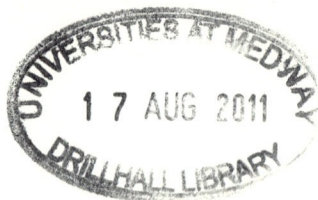

Published by the University of Minnesota Press
111 Third Avenue South, Suite 290
Minneapolis, MN 55401-2520
http://www.upress.umn.edu

Library of Congress Cataloging-in-Publication Data

Mobile urbanism: cities and policymaking in the global age / Eugene McCann and Kevin Ward, editors.
 p. cm. – (Globalization and community; v. 17)
 Includes bibliographical references and index.
 ISBN 978-0-8166-5628-8 (hc : alk. paper) – ISBN 978-0-8166-5629-5 (pb: alk. paper)
 1. Urban policy. 2. City planning. 3. Urbanization. I. McCann, Eugene. II. Ward, Kevin.

HT151.M559 2011
307.76–dc22   2010032726

Printed in the United States of America on acid-free paper

The University of Minnesota is an equal-opportunity educator and employer.

17 16 15 14 13 12 11   10 9 8 7 6 5 4 3 2 1

In memory of

Gary Gaile (1945–2009)

and

James McCann (1925–2010)

CONTENTS

FOREWORD

Allan Cochrane

Sometimes a book comes along that requires the reader to think about the world in different ways, to the extent that it is at once almost impossible to go back to thinking about it in the old way. Such books emerge at particular moments, capturing something fundamental about the times in which they are written. Once the ideas have been stated, it is difficult to believe that they were not always there, with the danger that they very quickly become taken for granted. But their insights provide the foundation on which whole new research agendas come to be built and new understandings developed.

This is one of those books.

Traditional approaches to urban politics have tended to focus on the local, sometimes through the filter of urban regimes, growth machines, or growth coalitions, sometimes through an institutional filter, whether driven by pluralist or elite theory, a concern with the detailed practices of local government, or an interest in community power (see, e.g., Davies and Imbroscio 2009). There have been debates about central–local relations, about the autonomy of urban government, and about the nature of urban governance—identifying and exploring shifts from managerialism to entrepreneurialism and back again in different forms. The policy focus has moved around, too, from local welfare state to local economic development, from urban regeneration to neighborhood renewal, from culture and creativity to environmentalism. But all of this has always been done with a focus on the urban as a local policy arena.

Sometimes, it is true, such discussions have been inflected by an acknowledgement that urban politics needs to be understood within a wider context, located within a wider framework. The significance of policy networks linking professional worlds stretching across government and administrative hierarchies has long been recognized, while the identification of national worlds of local government reflected in a range of organizational forms has also suggested the possibility of a less clearly delimited urban political sphere, one that is less obviously placed.

Unfortunately, it seems to have proved easier to draw these conclusions than to incorporate them into the practices of research into urban politics, in part because it is all too easy to slip back into the review of specific cases or wider generalization about the urban.

More recently, of course, the relationship between globalization and urban policy has been a focus of attention. Discussions of world cities and hierarchies of cities have tended to give them special status, however, with less attention given to the "worlding" of other urban spaces, including those that may be understood as "ordinary cities." Urban competitiveness in a global marketplace has become a shared concern of academic and policy literatures as debates have moved between emphasizing the need for successful cities to open themselves up to the market, to bemoaning the ideological hegemony of neoliberalism as a political force or, more positively, searching for the spaces to develop alternative visions at the urban level.

These various attempts to bring together the global and the local in the context of the city (sometimes even in a language of "glocalization") reflect the necessity of coming to terms with some of the more complex realities of urban governance and urban politics. They suggest that it is no longer possible to imagine the world through lenses that implicitly or explicitly locate things within nested scalar hierarchies. Instead, it is necessary to acknowledge the extent to which urban politics, by its very nature, incorporates actors and interests that are often implicitly and explicitly assumed to be located elsewhere. However, such an acknowledgment still leaves us grappling to understand what a politics that moves beyond such limitations might look like, how it is constructed, how—as this book helpfully poses the question—it is assembled.

The book goes beyond posing the question to beginning to provide some answers, by considering what it means to focus on the international or transnational politics of the urban and urban policy. In doing this, it goes far beyond dry notions of policy transfer as they have traditionally been understood—in more or less straightforward processes of borrowing and translating. Instead, the authors and editors explore ways in which global policy networks are fundamental to the construction of apparently local responses, while, at the same time, apparently global phenomena, globalized policies, only exist in particular, grounded, localized ways. The chapters highlight some of the tensions and some of the possibilities associated with these developments.

At the heart of the argument are two significant insights. The first is that urban policy and politics cannot be taken as given. It is possible neither to find some way of defining them as scientific objects of study nor to discover some formula that will fix them in place and explain their operation as flowing from some necessary set of relations. What matters is that they are part of an active and continuing process of production—they are produced, assembled in particular ways at particular times. It is this process that needs to be the focus of theoretical and empirical investigation. The second is that urban politics cannot just be approached through place (although its placing is important) but is also global or international in a much deeper sense—because of the extent to which it is based on systems of borrowing, reinterpreting, learning, and building networks. Indeed, from this perspective, it is precisely these connections that make it possible to understand and explore forms of urban politics as they are practiced in particular places.

Here, these issues are approached through a series of discretely organized chapters on substantive issues that nevertheless come together to produce a coherent whole. The book offers a mosaic of essays that implicitly engage, and sometimes debate (even disagree), with one another. They are fascinating cases in their own right but come together powerfully, not least because of the way in which the editors draw out crosscutting issues and frame them in their introductory and concluding chapters. One thing that makes this such a powerful book is how theory and practice, theory and evidence are so convincingly entwined throughout. The case studies are used to develop theoretical insights, and the theoretical insights are used to explore particular cases.

Everywhere is not the same as everywhere else, but everywhere borrows and reuses everything through particular practices, in ways that join it up with elsewhere all the time. In other words, elsewhere is right here as much as it is over there. At the core of the book is the recognition that far and near should no longer be taken for granted. This makes it possible to explore geographies of responsibility as they stretch across space, highlighting linkages that are otherwise rather too easy to avoid. It also makes it possible to explore the ways in which apparently distant phenomena can be drawn in by political actors to reinforce their position, to develop political initiatives, to resolve or generate political controversy, and to build political power and authority.

Reference

Davies, J., and D. Imbroscio, eds. 2009. *Theories of Urban Politics*. 2nd ed. London: Sage.

Urban Assemblages

Territories, Relations, Practices, and Power

Eugene McCann and Kevin Ward

> Increasingly, it would seem that there is little to be gained by talking about regional governance as a territorial arrangement when a number of the political elements assembled . . . are "parts" of elsewhere, representatives of professional authority, expertise, skills and interests drawn together to move forward varied agendas and programmes . . . There is . . . an *interplay* of forces where a range of actors mobilize, enrol, translate, channel, broker and bridge in ways that make different kinds of government possible.
> —Allen and Cochrane 2007, 1171; their emphasis

> The cumulative effect of a range of developments—the internationalization of consultancy firms; the broadening policy remits of transnational institutions; the formation of new policy networks around think tanks, governmental agencies and professional associations; and the growth of international conferencing and policy tourism—has been to proliferate, widen and lubricate channels of cross-border policy transfer.
> —Peck 2003, 228–29

The policy world seems to be one in constant motion. In a figurative sense, policymakers seem to be under increasing pressure to "get a move on"—to keep up with the latest trends and "hot" ideas that sweep into their offices, to convert those ideas into locally appropriate "solutions," and to "roll them out," thus making the most of them before the next fad arrives. As waves of innovation arrive more frequently, a concordant churning has been identified in urban policy, with new ideas and initiatives replacing old with increased regularity (Jessop and Peck 1998; Peck and Theodore 2001; Theodore and Peck 2001). Contemporary policymaking, at all scales, therefore involves the constant scanning (Gilbert 1999) of the policy landscape, via professional publications and reports, the media, Web sites, blogs, professional contacts, and word of mouth for ready-made, off-the-shelf policies and best practices that can be quickly applied locally.

It is in this context of "fast policy transfer" (Peck and Theodore 2001, 429) that figurative motion in the policy world becomes literal motion. Policy actors (a broadly defined category, including politicians, policy professionals, practitioners, activists, and consultants) act as transfer agents (Stone 2004), shuttling policies and knowledge about policies around the world through conferences, fact-finding study trips, consultancy work, and so on. These travels involve the transfer of policies from place to place (Dolowitz and Marsh 1996, 2000; Stone 1999, 2004; Evans 2004), which, in some cases, seem to diffuse with lightening speed: for example, welfare policies (Peck and Theodore 2001; Theodore and Peck 2001) and creative city policies (Florida 2002; Peck 2005). These travels and transfers involve local and national policymakers in networks that extend globally, bringing certain cities into conversation with each other (while pushing others farther apart). They create mental maps of "best cities" for policy that inform future strategies: Austin for quality of life and creativity (Florida 2002; McCann 2004b), Barcelona and Manchester for urban planning and regeneration (Gdaniec 2000; Bell and Binnie 2005; Monclús 2003; Peck and Ward 2002), Curitiba for environmental planning (Moore 2007), Portland for growth management (Calthorpe and Fulton 2001; Knaap and Nelson 1992), Porto Alegre for participatory budgeting and direct democracy (Baiocchi 2003). Thus, in a policy sense, as in other ways, cities are constituted through their relations with other places and scales (Massey 1991, 1993, 1999, 2005, 2007).

Yet, while motion and relationality define contemporary policymaking, this is, of course, only half the picture. Policies and policymaking are also intensely and fundamentally local, grounded, and territorial. The aforementioned examples confirm this point, since our ability to refer to complex approaches to vexing problems of urban governance through the use of a shorthand of city names indicates how tied certain policies are to specific places. There is a Barcelona model of urban regeneration, for example, which is contingent on the historical–geographical circumstances of that city and its relationship with other regional and national forms of decision making. While other cities might be encouraged to learn or to adopt the model, it is generally understood that, in doing so, adjustments will need to be made for it to work in those other locales. Similarly, it is understood that the success of participatory budgeting in Porto Alegre, for example, will not necessarily guarantee its successful adoption elsewhere. Furthermore, policy is fundamentally territorial in that it is tied up with a whole set of locally dependent interests, with those involved in growth coalitions being the most obvious examples (Logan and Molotch 1987; Cox and Mair 1988). As such, policymaking must be understood as both relational and territorial, as both in motion and simultaneously fixed, or embedded in place. The contradictory nature of policy should not, however, be seen as detrimental to its operation. Rather, the tension between policy as relational and dynamic, on the one hand, and fixed and territorial, on the other, is a productive one. It is a necessary tension that produces policy and places (cf. Harvey 1982, 1993).

It is in this context that, following Allen and Cochrane's comments in the epigraph, we deploy the notion of cities as *assemblages,* a rubric under which to frame the travels and transfers, political struggles, relational connections, and territorial fixities/mobilities brought together to constitute urbanism (Farías and Bender 2010; McFarlane 2011). The concept derives from Deleuze and Guattari's work and speaks not to the static arrangement of a set of parts, whether organized under some logic or collected randomly, but to "the *process* of arranging, organizing, fitting together . . . [where] an assemblage is a whole of some sort that expresses some identity and claims a territory" (Wise 2005, 77; his emphasis). More strongly, "assemblages create *territories*. Territories are more than just spaces: they have a stake, a claim . . . Territories are not fixed for all time, but are always being made and unmade, reterritorializing and deterritorializing. This constant making and unmaking process is the same with assemblages: they are always coming together and moving apart" (79; his emphasis). For Deleuze and Guattari (1987, 406; their emphasis) themselves, "an *assemblage* [is] every constellation of singularities and traits deducted from the flow—selected, organized, stratified—in such a way as to converge (consistency) artificially and naturally; an assemblage, in this sense, is a veritable invention."

This definition highlights three aspects of urban assemblages that will be explored in the chapters that follow. First, not only are assemblages and territories constituted by elements "deducted from the flow," but we would want to argue that these assemblages, in turn, shape, reorient, and reconstitute wider flows, thus continually reconfiguring geographies of territoriality and relationality. Second, assemblages, by definition, embody the sorts of tensions and allow us to overcome the sorts of easy analytical dichotomies—fixity/mobility, global/local—that we have already discussed. Collier and Ong, for instance (2005, 12) argue that "in relationship to 'the global,' the assemblage is not a 'locality' to which broader forces are counterposed. Nor is it the structural effect of such forces. The temporality of an assemblage is emergent. It does not always involve new forms, but forms that are shifting, in formation, or at stake." The aforementioned allusions to processes of "invention," "formation," and to what is at stake in any process of assemblage, such as the assemblage of urban policies, speak to a third aspect of urban assemblages: they are achievements with uneven consequences—they involve practices and politics. In theorizing the politics of regional governance, Allen and Cochrane (2007, 1171; emphasis added), put it this way: an (urban) region is "an assemblage of central, regional and local actors engaged in a complex set of political mobilizations at one point in time . . . all are part and parcel of a 'regional' assemblage of political power that is *defined by its practices*, not by some predetermined scalar arrangement of power" (see also Ward 2006; McCann 2011; Prince 2010; McFarlane 2011).

Our purpose is to explore the implications of these themes for our understanding of urban policy and to use the study of the "local globalness" of urban policy to inform the study of urban–global relations more generally. We will discuss how contemporary scholarship across the social sciences exhibits a remarkable convergence around questions of interscalar relations and around a conviction that specific cases of regulation, design, or policymaking, for example, must be understood in terms of how, where, why, and with what consequences urban actors assemble elements and resources from wider geographical fields and how the imperatives involved in these assembling and mobilizing practices may not be immediately evident at the scale of, or on the face of, the cases themselves. This approach, say Olds and Thrift (2005, 271), "makes . . . room for space. Assemblages will function quite differently, according to local circumstance, not because they are an overarching structure adapting its rules to the particular situation, but because these manifestations are what the assemblage consists of."

This is an important moment in which to consider global–urban assemblages. Ongoing discussions about the relationships between cities and global processes emphasize this point (McCann 2002, 2004a; Robinson 2002, 2006; Taylor 2004; Ward 2006), as do discussions about how best to conceptualize space—as networked, relational, and topological;

consisting of flows, relations, connections, disruptions, and folds; or as structured in ways where territorial fixings remain salient (Allen et al. 1998; Allen and Cochrane 2007; Amin 2002, 2007; Cox 1995, 2002; Jones 2009; Massey 2005; Murdoch and Marsden 1995). These discussions indicate that cities are important nodes in a "globalizing" world and that a focus on the practices that constitute cities as sites "of intersection between network topologies and territorial legacies" (Amin 2007, 103; chapter 2) is analytically crucial. Yet scholars still do not understand in a deep and detailed way *how* those involved in urban politics and policymaking act beyond their own cities to practice or perform urban globalness and to articulate their cities in the world. So, while we will outline a convergence of thought around the need for empirical detail on global political–economic relations, we will also suggest that the literature needs more empirical accounts of the struggles, practices, and representations that underpin urban–global relations and that assemble or territorialize global flows. The chapters in this book represent specifically urban examples of "a mode of inquiry that remains close to practices, whether through ethnography or careful technical analysis" (Collier and Ong 2005, 4) but that does not lose sight of the contexts and constraints within which these practices are located and by which they are channeled.

In the following section, we outline the convergence of work on scalar relations and, through a critical discussion of the "traditional" political science literature on policy transfer, we connect a relational–territorial approach to our specific empirical concerns. As Olds (2001, 9, citing Murdoch 1997, 334–35) puts it, "the 'role of the analyst,' is . . . 'to follow networks as they stretch through space and time, localizing and globalizing along the way.'" This is what the chapters in this book do. They draw on a range of empirical cases, including the recent London–Caracas pact, which exchanged oil for governance expertise (Massey, chapter 1); the emergence of collaborative planning and visioning strategies in a range of the world's cities, both the rich and the poor, including Johannesburg, Dar es Salaam, and London (Robinson, chapter 2); the historical and place-specific contexts of the globally popular creativity narrative in contemporary urban economic development policy, showing the connections and disconnects between London in the 1980s and Detroit in 2008 (Peck, chapter 3); the North American origins and British (re)territorialization of Business Improvement Districts (Ward, chapter 4); knowledge of elsewhere—Frankfurt, Zürich, and a therapeutic community outside Rimini—in the development and politics of urban drug policy in Vancouver (McCann, chapter 5); the complex global–urban geographies of response to the Severe Acute Respiratory Syndrome (SARS) outbreak, with a focus on Toronto, Singapore, and Hong Kong (Keil and Ali, chapter 6); and the urban implications of the globalization of airport management (McNeill, chapter 7). The authors pay close attention to (1) how urban policies are set in motion globally and how global circuits of policy knowledge and the transfer of policy models influence the governance of specific cities; and (2) how the assembling of policy and, by extension, of cities is a fundamentally territorialized and territorializing political process, contingent on specific historical–geographical circumstances.

Conceptualizing Global–Urban Connections:
Relationalities, Territorialities, Policies

In order to conceptualize global–urban connections as they happen in and through policymaking, it is necessary to account for policies' relational and territorial geographies and to address the related notion of policy mobilities.

The Relational and Territorial Geographies of Urban Policies

A great deal of critical geographical scholarship on cities examines the connections between urbanization and capitalism, the changing territorial forms of the state, and the production of new institutional arrangements for urban and regional governance, focusing on economic development and the "new urban politics" (Cox 1995; Cox and Mair 1988; Harvey 1989a, 1989b; Jessop 1998; Jonas and Wilson 1999; Lauria 1997; Logan and Molotch 1987; Peck 1995; Stone 1989; Ward 2000). Recent manifestations of this work entail the study of neoliberalization processes and their "actually existing" urban expressions (Brenner and Theodore 2002; Peck and Tickell 2002).

Neoliberalism promotes the extension of competitive market ideals to all aspects of life and, more specifically in relation to our purposes, to the operation of the state. Cities and the local state have been "entrepreneurialized" (Harvey 1989a; Leitner 1990), and the state has, through innovations like the New Public Management (Barzelay 2001), been reconceived as a facilitative, rather than regulatory, apparatus, behaving like a business to attract and support capital, rather than to promote welfare as understood under Keynesianism. As such, urban governance has been characterized by processes of downsizing, outsourcing, and privatization where services—from garbage collection, to policing, to policy expertise—are increasingly provided by the private sector and where urban governance frequently involves an extrospective, reflexive, and aggressive posture on the part of local elites and states . . . Today cities must actively—and responsively—scan the horizon for investment and promotion opportunities, monitoring "competitors" and emulating "best practice," lest they be left behind in this intensifying competitive struggle for the kinds of resources (public and private) that neoliberalism has helped make (more) mobile. (Peck and Tickell 2002, 47)

This context of competitive emulation produces, valorizes, mobilizes, and fundamentally depends on specific forms of what Larner (2002, 659) calls "post-welfarist" expertise and policy knowledge. The expertise of various think tanks, consultants, gurus, and mediators has become central to the day-to-day governance of cities and to their long-term aspirations for prosperity—a point illustrated in this volume by Peck's focus on creativity/cultural policies (chapter 3), Robinson's consideration of "city strategies" (chapter 2), and Ward's discussion of Business Improvement Districts (chapter 4).

Thus, the restructuring of the state at all levels and the reordering of linkages among state agencies, private business (including private policy experts), and communities of various forms has had impacts both on the "internal" character of cities and also on the "external" linkages among cities—nationally, regionally, and globally—and between urban policy actors and global circuits of policy knowledge (McCann 2011). Yet very little literature exists on *how*—through what practices, where, when, and by whom—urban policies are produced in global–relational context, are transferred and reproduced from place to place, and are negotiated politically in various locations. That said, a number of influential—although varied and not always entirely compatible—theorizations have sought to understand the tensions and power relations central to these global–urban connections. Four long-standing approaches, all of which act as contexts and reference points for the discussions in this volume, are (1) Harvey's (1982) conceptualization of the dialectic of fixity and mobility in capitalism and the implications of investment and disinvestment for urban built environments; (2) Massey's (1993, 2005) notion of a global sense of place in which specific places are understood to be open to and defined by situated combinations of flows of people, communications, responsibilities, and so forth, that extend far beyond specific locales; (3) the literature on spatial scale (Brenner 2001, 2004; Jonas 1994; Marston 2000; Smith 1993); and (4) the world/global cities literature with its focus on certain cities as powerful nodes in the networked geographies of finance capital (Brenner and Keil 2006; Taylor 2004).

These theorizations are complemented in some useful and interesting ways by two burgeoning literatures that also serve as resources and referents for a number of the chapters in this volume. The first is the "mobilities" approach (Cresswell 2001, 2006; Hannam et al. 2006; Sheller and Urry 2006a, 2006b). The purpose of this literature is to conceptualize the social content of movements of people and objects from place to place at various scales and the immobilities and moorings that underpin and challenge these dynamics. While this relatively recent literature is by no means coherent or unproblematic, a number of its insights inform Keil and Ali (chapter 6), McCann (chapter 5), and McNeill (chapter 7). A second, more established approach that has recently been engaged by geographers involves the neo-Foucauldian literature on governmentality (Rose and Miller 1992; Dean 1999; Rose 1999). Like the mobilities approach, geographers working on the role of governmental technologies and rationalities in shaping contemporary urban–global political economies have emphasized the seemingly mundane techniques, discourses, representations, and practices through which policy is "made up" (Larner 2000, 2003; Ward 2006), transferred, and enacted. For Robinson (chapter 2), this approach suggests that policies are always open to renegotiation and to potentially progressive outcomes in and through the ways they are practiced at the local level. For Ward (chapter 4), neoliberalism, in the guise of governable downtown business spaces, can only be fully understood through careful analysis of practices of comparison and translation and through the study of technologies like crime rates and other "performance indicators" (Larner and Le Heron 2002).

These literatures offer opportunities to conceptualize how cities are produced in relation to processes operating across wider geographical fields, while recognizing that urban localities simultaneously provide necessary basing points for those wider processes. Each suggests that there can be no easy separation between processes of territorialization and deterritorialization; between place-based and global–relational conceptualizations of contemporary political economies. As Beaumont and Nicholls (2007, 2559) argue, "territories do not come at the expense of extensive networks and flows but, rather, they are constituted by and contribute to these social networks." Hannam et al. (2006, 5) agree, "Mobilities cannot be described without attention to the necessary spatial, infrastructural and institutional moorings that configure and enable mobilities." Rather, according to Brenner (2004, 66), "the image of political-economic space as a complex, tangled mosaic of superimposed and interpenetrating nodes, levels, scales, and morphologies has become more [analytically] appropriate than the traditional Cartesian model of homogeneous, self-enclosed and contiguous blocks of territory." The tensions and crises implied by this "multiplex" (Amin and Graham 1997) and multiscale urban experience are objects of policymaking and politics. Harvey's (1989b) account of urban politics is particularly clear on this issue: while it is important to understand cities as always in a process of becoming (see also Berman 1982), social relations, state policy, and politics shape and are shaped by "structurally coherent" urban regions, or territories, which exist "in the midst of a maelstrom of forces that tend to undermine and disrupt" that very coherence (Harvey 1989b, 143).

Allen and Cochrane's discussion of (urban) regions resonates strongly with this viewpoint. The city is thus a social and political product that cannot be understood without reference to its relations with various other scales (Massey 1993, 2005). Yet to study how this social production *gets done*—"the actual *practices* of power" (Allen and Cochrane 2007, 1171; their emphasis)—involves analysis of a whole series of very specific and situated interactions, practices, performances, and negotiations. Allen and Cochrane argue that "some of this interplay takes place at arms length, mediated indirectly, some through relations of co-presence in a more distanciated fashion, and other forms of interaction are more direct in style, but together they amount to a more or less ordered assembly of institutional actors performing the 'region'" (Allen and Cochrane 2007, 1171), and again, by inference, the "city." Conceptualizing urban policymaking and politics in this way entails both the study of how urban actors manage and struggle over the "local" impacts of "global" flows and the analysis of how they engage in global circuits of policy knowledge that are produced in and through a "relational geography focused on networks and flows" (Olds 2001, 6).

These transfer agents seek, through this engagement, to take policy models from their own cities and promote them as "best practices" elsewhere or to tap into a global field of expertise to identify and "download" models of good policy. This process of territorializing and deterritorializing policy knowledge is highly political (McCann 2004b, 2008; Peck and Theodore 2001; Peck 2006; Ward 2006). "[Zones] of connectivity, centrality, and empowerment in some cases, and of disconnection, social exclusion and inaudibility" (Sheller and Urry 2006a, 210) are brought into being as struggles ensue over how policies get discursively framed as successes, while the insertion of new best practices from elsewhere into specific cities can empower some interests at the expense of others, putting alternative visions of the future outside the bounds of policy discussion. The construction of "models" of redevelopment and their circulation and reembedding in cities around the world can have profoundly disempowering consequences. However, this process of policy transfer can also spur contests within cities where activists question the "preapproved" credentials of newly imported policy models or where activists are motivated to perform their own global scans to find alternative policies as part of what Purcell (2008, 153) calls "fast resistance transfer."

From Policy Transfer . . . to Mobile Policies

How might we think specifically about the assemblage of *urban policies* and *the urban* from a relational–territorial perspective? An appropriate point of departure is to consider the already existing political science literature on policy transfer, which studies how policies are learned from one context and moved to another with the hope of similar results. In one sense, this is a literature that is all about global relations and territories. It developed in the 1990s as "the scope and intensity of policy transfer activity . . . increased significantly" (Evans 2004, 1). This voluminous literature, while internally differentiated and heterogeneous in some respects, also shares some common features. It focuses on modeling how transfer works, creating typologies of "transfer agents" (Stone 2004), and identifying conditions under which transfer leads to successful or unsuccessful policy outcomes in the new location (Dolowitz and Marsh 2000; Evans 2004; Hulme 2005; Jones and Newburn 2006; Stone 1996, 1999; Walker 1999). So it is not without its insights.

Yet while this literature is certainly about global relations and territories, it has not involved an engagement with the full range of social territoriality. It is limited in its definition of the agents involved in transfer, focusing largely on national and international elites, usually working in formal institutions. It focuses solely on national territories—transfer among nations or among localities with single nations—without considering the possibility, or actuality, of intercity transfers that transcend national boundaries, connecting cities globally (an exception is Hoyt 2004). Furthermore, it tends *not* to consider transfer as a sociospatial process in which policies are changed as they travel (Peck and Theodore 2001; McCann 2008, 2011).

These limits to the "traditional" policy transfer literature offer a series of opportunities for further theorization from a number of perspectives that understand, often in different ways, transfer as a global–relational, social, and spatial process that interconnects and constitutes cities (Cook 2008; Guggenheim and Söderström 2010; Healey and Upton 2010; McCann 2004b, 2008, 2010; Peck and Theodore 2001; Peck 2003; Prince 2010; Wacquant 1999; Ward 2006, 2007). For Wacquant (1999, 321), the aim should be "to constitute, link by link, the long chain of institutions, agents and discursive supports" that constitute the current historical period while Peck (2003, 229) suggests the challenge is

> to develop adequate conceptualizations and robust empirical assessments of policies "in motion," including descriptions of the circulatory systems that connect and interpenetrate "local" policy regimes. This calls for an analytical shift of sorts, away from the traditional method of focusing on the internal characteristics of different regimes—qua taxonomically defined "systems"—and towards the transnational and translocal constitution of institutional relations, governmental hierarchies and policy networks.

Larner (2003, 510), taking a governmentality approach, also advocates a move in the same intellectual direction, toward a "more careful tracing of the intellectual, policy, and practitioner networks that underpin the global expansion of neoliberal ideas, and their subsequent manifestation in government policies and programmes." Explicitly interested in understanding both how and why governing practices and expertise are moved from one place to another, she advocates the "detailed tracings" of social practices, relations, and embeddings (Prince 2010). For example, her study of the global call center and banking industries and the place of New Zealand in the globalization of these economic activities shows the value of the detailed rendering of what might be seen as the banal or mundane practices of various actors who, individually and collectively, play an important role in constituting neoliberal globalization (Larner 2001).

Much of the mobilities work attempts to understand the details of a particular form of mobility, or a specific infrastructure that facilitates or channels mobilities, in reference to wider processes and contexts:

> [It] problematizes both "sedentarist" approaches in the social sciences that treat place, stability, and dwelling as a natural steady-state, and "deterritorialized" approaches that posit a new "grand narrative" of mobility, fluidity or liquidity as a pervasive condition of postmodernity or globalization . . . It is a part of a broader theoretical project aimed at going beyond the imagery of "terrains" as spatially fixed geographical containers for social processes, and calling into question scalar logics such as local/global as descriptors of regional extent . . . (Hannam et al. 2006, 5; see also Sheller and Urry 2006a, 2006b)

For us, the language of the mobilities approach is a useful frame for our discussion of mobile policies because it emphasizes the social and the scalar, the fixed and mobile character of policies. We utilize "mobilities" in the sense that people, frequently working in institutions, mobilize objects and ideas to serve particular interests and with particular material consequences.

We can, then, see convergences among scholars about the need to be alive to both the *why* and the *how* of policy transfer. This demands that we pay attention to how—through "ordinary" and "extraordinary" activities—policies are made mobile (and immobile), to why this occurs, and to relationships between these mobilities and the sociospatial (re) structuring of cities. The question remains, How might we best frame these sorts of empirical discussions? Should we understand contemporary policymaking as primarily about territory, as primarily about relationality, or in terms of a both/and logic that recognizes that contemporary "global restructuring has entailed neither the absolute territorialization of societies, economies, or cultures onto a global scale, nor their complete deterritorialization into a supraterritorial, distanceless, placeless, or borderless space of flows" (Brenner 2004, 64)? The authors assembled in this book take the latter position, but, as one might expect, they develop their detailed case studies in varied, although overlapping, ways.

The Chapters

In their own ways, the following chapters address a number of broader issues that we will elaborate on here since, for us, they refer directly to the process of assembling urbanism through policymaking and policy mobilizing. First, the collection develops a conceptualization of urban policymaking and place making as an assemblage of "territorial" and "relational" geographies, recognizing the material reality of each and the productive tensions between them, without affording either of them preeminence. The authors show that cities are assembled by the situated practices and imaginations of actors who are continually attracting, managing, promoting, and resisting global flows. Following Olds (2001, 8), we advocate "a relational geography that recognizes the contingent, historically specific, uneven, and dispersed nature of material and non-material flows." Second, most of the authors offer more or less explicit critiques of the existing literature on policy transfer. They demonstrate a broad understanding of "transfer agents," take seriously the transfer of interurban, transnational knowledges, policies, practices, and urban forms, and they understand "transfer" as a sociospatial, power-laden process in which policies are subject to change and struggle as they are moved.

Third, the chapters offer varied insights into the methodological implications of analyzing urban assemblages and global circuits of knowledge from a relational and practice-oriented perspective. There is value in paying attention to how various spaces are brought into being during the journey of a policy or program or in attempts to manage the movement of global flows in and through cities. Like Burawoy (2000), they underscore the need to conduct "global ethnographies" that attend concurrently to particular contexts and to global forces, connections, and imaginaries. For the contributors to this volume, following policies, policymakers, experts, and regulators through specific urban–global spaces involves a set of largely qualitative and ethnographic methods. There is a strong emphasis on interviews with key actors and on analyses of the discourses revealed in these interviews and in various documents and public statements. Insights gained from these sources are also often supported by knowledge of the cases that stems from participation and observation. These methods are particularly appropriate since, as Collier and Ong (2005, 14; emphasis added) put it, a "function of the study of assemblages is to gain analytical and critical insight into global forms by examining how actors *reflect upon them and call them into question*." Furthermore, Robinson (chapter 2), in particular, suggests that it is incumbent on researchers also to reflect continually on the ethical and political implications of their methods and analyses: the study of knowledge in motion must not assume that concepts developed in Europe and in North America can or should be used to analyze case studies elsewhere (Pollard et al. 2009; Ward 2010). Fourth, the methodological and analytical approach demonstrated here has, at its core, a sensitivity to the mutual constitution of structure and agency (Olds and Thrift 2005): time and again, the chapters show that key individuals make a difference. This is not done under terms of their making, however. Rather the chapters also indicate, in their own ways, how a set of macro supply and demand contexts frame policy mobilities and empower certain "idea brokers" (Smith 1991) or mediators (Osborne 2004). Some, more than others, are likely to have their ideas and policies made mobile, and they are, in turn, most likely to stamp their authority on emergent urban assemblages/territories. The policymaking landscape is an uneven one, produced through past, and contributing to future, unevenness. Yet, as some of the chapters reveal, subaltern or oppositional groups as well as governments with ideologies that differ from the global neoliberal norm, can inhabit and use the same global circuits of policy knowledge to develop alternative assemblages of policy and power.

It is to the urban–global production of just this sort of counterhegemonic globalization that Doreen Massey speaks (chapter 1). She discusses the agreement between the government of Venezuela and London's city government, which lasted from 2006 to 2008 and entailed both cultural exchanges between London and Caracas and, more famously, the trade of cheap oil from Venezuela in exchange for London's expertise on matters of urban public administration. This exchange and its "politics of place beyond place" emphasize the mutual constitution of relationality and territoriality, Massey argues, and it illuminates the potential of counterhegemonic projects and alliances that create interurban relations on the basis of equal exchange and without an ideology of competition. Yet this counterhegemonic relationality is entwined with both cities' links to two totems of the contemporary global economy—finance and oil—and Massey's explores the tensions and complexities of this context as the cities seek to participate in the local production of the global.

The cautious optimism of Massey's account and her attention to transfers that link wealthier and poorer cities are also evident in Jennifer Robinson's contribution (chapter 2), which sets an agenda for conceptualizing the global proliferation of "city strategies," or comprehensive, collaborative visioning exercises. She identifies the various forms and locations of these strategies, offering examples from Africa, Latin America, and Europe, and suggests, first, that they offer the opportunity to develop a postcolonial and topological conceptualization of urban–global relations, which refuses the notion that best policy practices can be traced to Western points of origin. Second, she advocates a neo-Foucauldian approach to the institutional alliances and power relations involved in mobilizing and territorializing policy models that is sensitive to various forms of power involved in the development and circulation of policies, thus allowing her to identify the potential within city strategies for alternatives to neoliberalism and progressive strategies that might be developed locally and also circulated more widely. The likelihood of these strategies emerging depends, she argues, on the contexts—local, national, global—in which they are developed and speaks to the "indeterminacy" of policy circulation, a phrase that resonates with the notion of assemblage discussed earlier.

While Massey and Robinson speak to the possibilities for anti-neoliberal political practices that can emerge in and through the same geographies of policy circulation as neoliberal policy transfers, neoliberal ideologies continue to be important frames for contemporary urbanism. Jamie Peck (chapter 3) and Kevin Ward (chapter 4) address two important policy models—cultural–creativity strategies and Business Improvement Districts—that have become globally mobile in recent years and, through their globality, have shaped local economic development strategy in numerous cities. Peck echoes Robinson's arguments about the importance of context in his analysis of the contemporary policy fad that advocates for cities to be "creative." To illustrate his point, he broadens the contemporary focus on the "creative class" and sets its production and circulation in a spatial and historical context. He contrasts an earlier case of urban economic development actors, enlisting culture and creativity to the task of improving cities. This case was in 1980s Britain, with a primary focus on London. He identifies the (dis)connections in politics, rhetoric, and outcomes between this strategy and the contemporary creative cities approach that has spread across the world and is seen as the best hope for cities like Detroit. For him, each attempt to combine urban creativity and development in policymaking must be seen to be "organically embedded in distinctive urban political cultures" that vary both spatially and historically. This discussion emphasizes that the English cultural strategies of the 1980s cannot be seen as an earlier point on a smooth trajectory that has now brought us the "creative city" movement. Rather, there are many crucial differences between the two approaches. These distinctions highlight the specifically neoliberal character of the contemporary creative policy fix while also emphasizing the need for studies of circuits of policy knowledge to understand mobile policies as conditioned by the specific contexts (or "policy ecologies") in which they are produced and consumed.

Similar concerns with globally popular urban development strategies and the historical and spatial cultures that have shaped them permeate Kevin Ward's discussion of Business Improvement Districts (BIDs) (chapter 4). These governance arrangements allow businesses to levy extra taxes to fund private, district-specific services and improvements, including advertising, private security, and trash collection and are both lauded and criticized for their appeal to the logic and "flexibility" of the private sector in urban governance. Ward shows how this model has deep and diverse roots in large North American cities and, using the example of England, shows how it has been territorialized by a range of transfer agents in specific ways in a variety of cities, including rather small and, at first glance, apparently "nonglobal" ones. What Ward shows is that BIDs, like all assemblages, are inventions or achievements that create their territories just as they are tied to complex policy circulation systems. Applying the general body of BID technopolitical expertise to specific cities is, as Ward details, a question of active translation and political persuasion in which local actors are persuaded of the local applicability of policies that emerged from very different contexts.

Mobilization, translation, and political persuasion are central to the assemblage of cities and urban policies, not only in terms of urban planning and economic development (although the first four chapters show

how important these elements are). Urban actors also engage with the global circulation of knowledge and regulation in other social contexts, including the governance of urban health. Eugene McCann (chapter 5) and Roger Keil and Harris Ali (chapter 6) provide differing but connected insights in this regard. McCann focuses on the politics of urban drug policy in Vancouver over the past fifteen years, when drug use and addiction have come to be officially understood as primarily public health, rather than criminal, concerns, and a new city drug strategy has produced, among other initiatives, a legal, supervised facility for users to inject illicit drugs in relative safety. This policy developed after harm reduction models from German and Swiss cities were learned, transferred, and modified for the Canadian context in a highly political process where knowledge of elsewhere was presented in careful translation to be persuasive. A decade later, harm reduction policies receive strong public support but still remain the object of political debate as opponents and skeptics invoke alternative models, such as an Italian abstinence-based approach, as policy correctives or replacements. McCann suggests that urban drug policy can be seen as an assemblage of "parts of elsewhere," that urban politics involves the strategic deployment of knowledge of elsewhere more frequently than our analyses acknowledge, and that any analysis of the politics of policy transfer must consider longer-term historical contexts in which knowledge of elsewhere resonates through urban political debates.

Keil and Ali (chapter 6) also address the relational–territorial aspects of how urban public health is governed. For them, the focus is the 2003 SARS outbreak that tied certain cities—Toronto, Hong Kong, Singapore— together in a network shaped by the rapid global circulation of an emerging infectious disease through air travel and by attempts to quickly share knowledge about how to halt its spread. SARS both revealed and reordered common understandings of the territorial integrity of sovereign nation-states in terms of health and highlighted the role of the local scale in health governance. The outbreak involved conflicts between the World Health Organization and national health authorities, markedly different response strategies in different countries, depending on the relationship between national and local authorities, and tensions between traditional understandings of the city as a closed, contained entity (on which public health strategies like quarantine are based) and contemporary understandings of global cities as networked and porous. As Ali and Keil note, traditional territorial containment measures still proved successful in response to SARS, even if its rapid spread and scientists' swift development of an understanding of its character were both products of global interconnectedness and even as the experience of dealing with it has led to rethinking the traditional scales and territories of health management. Territoriality and relationality were very much two sides of the same coin in this case.

The SARS outbreak has become associated with air travel and airports, not only in terms of the rapid mobility they afford, but also, as Keil and Ali suggest, in terms of the terminal infrastructures and procedures that were employed to limit its spread (e.g., heat sensors intended to detect asymptomatic carriers of the virus as they emerged from planes). Indeed, air mobilities and the fixed infrastructures that actualize them are crucial to the travels of all the forms of knowledge and policy discussed in this book, because policymakers, consultants, and activists frequently travel to see best practices firsthand. Therefore, we end the book with Donald McNeill's political economy of the global airport form, its governance, and its links to the city (chapter 7). McNeill provides a detailed discussion of how airports have globalized and how they are run now by global corporations, indeed, how they represent a form of mobile–territorializing management knowledge through which those who manage a single airport, like Schiphol, have mobilized their model and now govern many other airports on similar principles. A point running through the chapter is that airports are an urbanized form; they are not disembedded, purely relational spaces. They are, instead, very much territorialized. This is illustrated through discussions of airports' ambivalent relationship to their surrounding urban regions in the struggle over land use and transportation, on the one hand, and their role in the entrepreneurial efforts of city growth coalitions, on the other.

Jessop, Brenner, and Jones (2008, 391) warn against analytical perspectives that seek to theorize sociospatial relations but "fall into the trap of conflating a part (territory, place, scale, or networks) with the whole (the totality of sociospatial organization) whether due to conceptual imprecision, an overly narrow analytical focus, or the embrace of an untenable ontological (quasi-)reductionism." The chapters in this book chart a course through such dangers by drawing innovatively on a range of intersecting theories and applying them to carefully rendered empirical examples that illustrate precisely *how* relationality and territoriality are entwined and *how* urbanism is mobilized and territorialized—assembled—through policy practice.

References

Allen, J., and A. Cochrane. 2007. "Beyond the Territorial Fix: Regional Assemblages, Politics, and Power." *Regional Studies* 41:1161–75.

Allen, J., D. Massey, and A. Cochrane. 1998. *Re-thinking the Region.* London: Routledge.

Amin, A. 2002. "Spatialities of Globalization." *Environment and Planning A* 34:385–99.

———. 2007. "Re-thinking the Urban Social." *City* 11, no. 1: 100–114.

Amin, A., and S. Graham. 1997. "The Ordinary City." *Transactions of the Institute of British Geographers* 22:411–29.

Baiocchi, G. 2003. "Participation, Activism, and Politics: The Porto Alegre Experiment." In *Deepening Democracy: Institutional Innovations in Empowered Participatory Governance*, edited by A. Fung, E. O. Wright, R. N. Abers, and R. Abers, 45–76. London: Verso.

Barzelay, M. 2001. *The New Public Management: Improving Research and Policy Dialogue.* Berkeley: University of California Press.

Beaumont, J., and W. Nicholls. 2007. "Between Relationality and Territoriality: Investigating the Geographies of Justice Movements in The Netherlands and the United States." *Environment and Planning A* 39:2554–74.

Bell, J., and J. Binnie. 2005. "What's Eating Manchester? Gastro-culture and Urban Regeneration." *Architectural Design* 75:78–85.

Berman, M. 1982. *All That Is Solid Melts into Air: The Experience of Modernity.* New York: Penguin Books.

Brenner, N. 2001. "The Limits to Scale? Methodological Reflections on Scalar Structuration." *Progress in Human Geography* 15:525–48.

———. 2004. *New State Spaces: Urban Governance and the Rescaling of Statehood.* Oxford: Oxford University Press.

Brenner, N., and R. Keil. 2006. *The Global Cities Reader.* London: Routledge.

Brenner, N., and N. Theodore, eds. 2002. *Spaces of Neoliberalism: Urban Restructuring in North America and Western Europe.* London: Blackwell.

Burawoy, M. 2000. "Introduction: Reaching for the Global." In *Ethnography Unbound: Power and Resistance in the Modern Metropolis,* edited by M. Burawoy, J. A. Blum, S. George, Z. Gille, T. Gowan, L. Haney, M. Klawiter, S. H. Lopez, S. Ó Riain, and M. Thayer, 1–40. Berkeley: University of California Press.

Calthorpe, P., and W. B. Fulton. 2001. *The Regional City: Planning for the End of Sprawl.* Washington, D.C.: Island Press.

Collier, S. J., and A. Ong. 2005. "Global Assemblages: Anthropological Problems." In *Global Assemblages: Technology, Politics, and Ethics as Anthropological Problems,* edited by A. Ong and S. J. Collier, 3–21. Malden: Blackwell.

Cook, I. R. 2008. "Mobilising Urban Policies: The Policy Transfer of U.S. Business Improvement Districts to England and Wales." *Urban Studies* 444:773–95.

Cox, K. R. 1995. "Globalisation, Competition and the Politics of Local Economic Development." *Urban Studies* 32:213–24.

———. 2002. *Political Geography: Territory, State and Society.* Oxford: Blackwell.

Cox, K. R., and A. Mair. 1988. "Locality and Community in the Politics of Local Economic Development." *Annals of the Association of American Geographers* 78:307–25.

Cresswell, T. 2006. *On the Move: Mobility in the Modern Western World.* New York: Routledge.

———, ed. 2001. "Mobilities." Special issue, *New Formations* 43.

Dean, M. 1999. *Governmentality: Power and Rule in Modern Society.* Thousand Oaks, Calif.: Sage.

Deleuze, G., and F. Guattari. 1987. *A Thousand Plateaus: Capitalism and Schizophrenia.* Minneapolis: University of Minnesota Press.

Dolowitz, D., and D. Marsh. 1996. "Who Learns from Whom: A Review of the Policy Transfer Literature." *Political Studies* 44:342–57.

———. 2000. "Learning from Abroad: The Role of Policy Transfer in Contemporary Policy Making." *Governance* 13:5–24.

Evans, M., ed. 2004. *Policy Transfer in Global Perspective.* Aldershot: Ashgate.

Farías, I., and Bender, T., eds. 2010. *Urban Assemblages: How Actor Network Theory Changes Urban Studies.* Routledge: London.

Florida, R. 2002. *The Rise of the Creative Class: And How It's Transforming Work, Leisure, Community and Everyday Life.* New York: Basic Books.

Gdaniec, C. 2000. "Cultural Industries, Information Technology and the Regeneration of Post-Industrial Urban Landscapes. Poblenou in Barcelona—a Virtual City?" *Geojournal* 50:379–87.

Gilbert, A. 1999. "'Scan Globally. Reinvent Locally': Reflecting on the Origins of South Africa's Capital Housing Subsidy Policy." *Urban Studies* 39:1911–33.

Guggenheim, M., and O. Söderström. 2010. *Re-shaping Cities: How Global Mobility Transforms Architecture and Urban Form.* New York: Routledge.

Hannam, K., M. Sheller, and J. Urry. 2006. Editorial: Mobilities, Immobilities and Moorings. *Mobilities* 1:1–22.

Harvey, D. 1982. *The Limits to Capital.* Chicago: University of Chicago Press.

———. 1989a. "From Managerialism to Entrepreneurialism—the Transformation in Urban Governance in Late Capitalism." *Geografiska Annaler Series B* 71:3–17.

———. 1989b. *The Urban Experience.* Baltimore: Johns Hopkins University Press.

———. 1993. "From Space to Place and Back Again: Reflections on the Condition of Postmodernity." In *Mapping the Futures: Local Cultures, Global Change,* edited by J. Bird, B. Curtis, T. Putman, G. Robertson, and L. Tickner, 3–29. New York: Routledge.

Healey, P., and Upton, R., eds. 2010. *Crossing Borders: International Exchange and Planning Practices.* Routledge: London.

Hoyt, L. 2004. "Collecting Private Funds for Safer Public Spaces: An Empirical Examination of the Business Improvement Districts Concept." *Environment and Planning B* 31:367–380.

Hulme, R. 2005. "Policy Transfer and the Internationalization of Social Policy." *Social Policy and Society* 4:417–25.

Jessop, B. 1998. "The Narrative of Enterprise and the Enterprise of Narrative: Place Marketing and the Entrepreneurial City." In *The Entrepreneurial City: Geographies of Politics, Regimes and Representation,* edited by T. Hall and P. Hubbard, 77–99. Chichester: John Wiley and Sons.

Jessop, B., and J. Peck. 1998. "Fast Policy/Local Discipline: The Politics of Time and Scale in the Neoliberal Workfare Offensive." Mimeograph, Department of Sociology, Lancaster University.

Jessop, B., N. Brenner, and M. R. Jones. 2008. "Theorizing Sociospatial Relations." *Environment and Planning D: Society and Space* 26:389–401.

Jonas, A. E. G. 1994. "The Scale Politics of Spatiality." *Environment and Planning D: Society and Space* 12:257–64.

Jonas, A. E. G., and D. Wilson., eds. 1999. *The Urban Growth Machine: Critical Perspectives, Two Decades Later.* Albany: State University of New York Press.

Jones, M. 2009. "Phase Space: Geography, Relational Thinking and Beyond." *Progress in Human Geography* 33, no. 4: 487–506.

Jones, T., and T. Newburn. 2006. *Policy Transfer and Criminal Justice.* Milton Keynes: Open University Press.

Knaap, G., and A. Nelson. 1992. *The Regulated Landscape: Lessons on State Land Use Planning from Oregon.* Cambridge, Mass.: Lincoln Institute of Land Policy.

Larner, W. 2000. "Neo-liberalism: Policy, Ideology, Governmentality." *Studies in Political Economy* 63:5–25.

———. 2001. "Governing Globalisation: The New Zealand Call Centre Attraction Initiative." *Environment and Planning A* 33:297–312.

———. 2002. "Globalization, Governmentality, and Expertise: Creating a Call Centre Labour Force." *Review of International Political Economy* 9, no. 4: 650–74.

———. 2003. "Guest Editorial: Neoliberalism?" *Environment and Planning D: Society and Space* 21:508–12.

Larner, W., and R. Le Heron. 2002. "From Economic Globalization to Globalizing Economic Processes: Towards Post-Structural Political Economies." *Geoforum* 33:415–19.

Lauria, M., ed. 1997. *Reconstructing Urban Regime Theory.* Thousand Oaks, Calif.: Sage.

Leitner, H. 1990. "Cities in Pursuit of Economic Growth: The Local State as Entrepreneur." *Political Geography Quarterly* 9, no. 2: 146–70.

Logan, J., and H. Molotch. 1987. *Urban Fortunes: The Political Economy of Place.* Berkeley: University of California Press.

Marston, S. A. 2000. "The Social Construction of Scale." *Progress in Human Geography* 24, no. 2: 219–42.

Massey, D. 1991. "A Global Sense of Place." *Marxism Today*, June, 24–29.

———. 1993. "Power-Geometry and a Progressive Sense of Place." In *Mapping the Futures: Local Cultures, Global Change*, edited by J. Bird, B. Curtis, T. Putman, G. Robertson, and L. Tickner, 59–69. New York: Routledge.

———. 1999. "Imagining Globalisation: Power-Geometries of Space-Time." In *Global Futures: Migration, Environment, and Globalisation*, edited by A. Brah, M. Hickman, and M. MacanGhaill, 27–44. Basingstoke, UK: St. Martin's Press.

———. 2005. *For Space*. Sage: London.

———. 2007. *World City*. Polity Press: Cambridge.

McCann, E. J. 2002. "The Urban as an Object of Study in Global Cities Literatures: Representational Practices and Conceptions of Place and Scale." In *Geographies of Power: Placing Scale*, edited by A. Herod and M. W. Wright, 61–84. Cambridge, Mass.: Blackwell.

———. 2004a. "Urban Political Economy beyond the 'Global' City." *Urban Studies* 41:2315–33.

———. 2004b. "'Best places': Interurban Competition, Quality of Life and Popular Media Discourse." *Urban Studies* 41:1909–29.

———. 2008. "Expertise, Truth, and Urban Policy Mobilities: Global Circuits of Knowledge in the Development of Vancouver, Canada's 'Four Pillar' Drug Strategy." *Environment and Planning A* 40, no. 4: 885–904.

———. 2010. "Urban Policy Mobilities and Global Circuits of Knowledge: Toward a Research Agenda." *Annals of the Association of American Geographers* 101:107–30.

McFarlane, C. 2011. *Learning the City: Translocal Assemblages and Urban Politics*. Oxford: Wiley-Blackwell.

Monclús, F. J. 2003. "The Barcelona Model: And an Original Formula? From 'Reconstruction' to Strategic Urban Projects (1979–2004)." *Planning Perspectives* 18:399–421.

Moore, S. A. 2007. *Alternative Routes to the Sustainable City: Austin, Curitiba, and Frankfurt*. Plymouth: Lexington Books.

Murdoch, J. 1997. "Towards a Geography of Heterogeneous Association." *Progress in Human Geography* 21:321–37.

Murdoch, J., and T. Marsden. 1995. "The Spatialization of Politics: Local and National Actor Spaces in Environmental Conflict." *Transactions of the Institute of British Geographers* 20:368–80.

Olds, K. 2001. *Globalization and Urban Change: Capital, Culture, and Pacific Rim Mega-Projects*. Oxford: Oxford University Press.

Olds, K., and N. Thrift. 2005. "Cultures on the Brink: Reengineering the Soul of Capitalism—on a Global Scale." In *Global Assemblages: Technology, Politics, and Ethics as Anthropological Problems*, edited by A. Ong and S. J. Collier, 270–90. Malden, Mass.: Blackwell.

Osborne, T. 2004. "On Mediators: Intellectuals and the Ideas Trade in the Knowledge Society." *Economy and Society* 3, no. 4: 430–47.

Peck, J. 1995. "Moving and Shaking: Business Elites, State Localism, and Urban Privatism." *Progress in Human Geography* 19:16–46.

———. 2003. "Geography and Public Policy: Mapping the Penal State." *Progress in Human Geography* 27:222–32.

———. 2005. "Struggling with the Creative Class." *International Journal of Urban and Regional Research* 24:740–70.

———. 2006. "Liberating the City: Between New York and New Orleans." *Urban Geography* 27:681–723.

Peck, J., and N. Theodore. 2001. "Exporting Workfare/Importing Welfare-to-Work: Exploring the Politics of Third Way Policy Transfer." *Political Geography* 20:427–60.

Peck, J., and A. Tickell. 2002. "Neoliberalizing Space." In *Spaces of Neoliberalism*, edited by N. Brenner and N. Theodore, 33–57. Malden, Mass.: Blackwell.

Peck, J., and K. Ward, eds. 2002. *City of Revolution: Restructuring Manchester.* Manchester: Manchester University Press.

Pollard, J., C. McEwan, N. Laurie, and A. Stenning. 2009. "Economic Geography under Postcolonial Scrutiny." *Transactions of the Institute of British Geographers* 34, no. 2: 137–42.

Prince, R. 2010. "Policy Transfer as Policy Assemblage: Making Policy for the Creative Industries in New Zealand." *Environment and Planning A* 42, no. 2: 169–86.

Purcell, M. 2008. *Recapturing Democracy.* New York: Routledge.

Robinson, J. 2002. "Global and World Cities: A View from Off the Map." *International Journal of Urban and Regional Research* 26:531–54.

———. 2006. *Ordinary Cities: Between Modernity and Development.* London: Routledge.

Rose, N. 1999. *Powers of Freedom: Reframing Political Thought.* Cambridge: Cambridge University Press.

Rose, N., and P. Miller. 1992. "Political Power beyond the State: Problematics of Government." *British Journal of Sociology* 43, no. 2: 172–205.

Sheller, M., and J. Urry. 2006a. "The New Mobilities Paradigm." *Environment and Planning D: Society and Space* 38:207–26.

———, eds. 2006b. *Mobile Technologies of the City.* Aldershot: Ashgate.

Smith, J. 1991. *The Idea Brokers: The Rise of Think Tanks and the Rise of the Policy Elite.* New York: Free Press.

Smith, N. 1993. "Homeless/Global: Scaling Places." In *Mapping the Futures: Local Cultures, Global Change*, edited by J. Bird, B. Curtis, T. Putman, G. Robertson, and L. Tickner, 87–119. New York: Routledge.

Stone, C. 1989. *Regime Politics: Governing Atlanta, 1946–1988.* Lawrence: University of Kansas Press.

Stone, D. 1996. *Capturing the Political Imagination: Think Tanks and the Policy Process.* London: Frank Cass.

———. 1999. "Learning Lessons and Transferring Policy across Time, Space, and Disciplines." *Politics* 19:51–59.

———. 2004. "Transfer Agents and Global Networks in the 'Transnationalisation' of Policy." *Journal of European Public Policy* 11:545–66.

Taylor, P. 2004. *World City Network: A Global Urban Analysis.* London: Routledge.

Theodore, N., and J. Peck. 2001. "Searching for 'Best Practice' in Welfare-to-Work: The Means, the Method and the Message." *Policy and Politics* 29:85–98.

Wacquant, L. 1999. "How Penal Common Sense Comes to Europeans: Notes on the Transatlantic Diffusion of the Neoliberal Doxa." *European Societies* 1:319–52.

Walker, R. 1999. "The Americanization of British Welfare: A Case Study of Policy Transfer." *International Journal of Health Services* 29:679–97.

Ward, K. 2000. "Critique in Search of a Corpus: Re-visiting Governance and Re-interpreting Urban Politics." *Transactions of the Institute of the British Geographers* 25:169–85.

———. 2006. "'Policies in Motion,' Urban Management and State Restructuring: The Trans-local Expansion of Business Improvement Districts." *International Journal of Urban and Regional Research* 30:54–75.

———. 2007. "Business Improvement Districts. Policy Origins: Mobile Policies and Urban Liveability." *Geography Compass* 2:657–72.

Ward, K. 2010. "Entrepreneurial Urbanism and Business Improvement Districts in the State of Wisconsin: A Cosmopolitan Critique." *Annals of the Association of American Geographers* 100, no. 5:1177–96.

Wise, J. M. 2005. "Assemblage." In *Gilles Deleuze: Key Concepts,* edited by C. J. Stivale, 77–87. Montreal and Kingston: McGill and Queen's University Press.

CHAPTER ONE

A Counterhegemonic Relationality of Place

Doreen Massey

In the spring of 2008, New Labour was routed in elections throughout the United Kingdom. Its project to stitch together a full-blown commitment to privatization, deregulation, and the financial City, on the one hand, and an obeisance to social democratic concerns for the welfare state and for the poor, on the other, had run into the ground. In all parts of the country, New Labour lost seats on local councils. The swing against them was enormous, and votes went to the only other party available, the Conservatives.

It was a comprehensive defeat. In the tidal wave of rejection, one city held out better than anywhere else. In London, the electoral battle was seriously local (it was about the politics of the city itself) as well as national. Two charismatic figures faced each other. On the left was incumbent mayor Ken Livingstone, who had been in power since 2000 and had already been reelected once. In the 1980s, Livingstone had led the Greater London Council, which, as part of the "new urban Left" that rose up against the politics of Margaret Thatcher, had been summarily abolished. When New Labour regained power nationally, in 1997, they reestablished London government, though in a weaker and more corseted form. "Red Ken" was reelected. His opponent in the elections of 2008 was indubitably of the right: upper-class Conservative Boris Johnson, whose jokey personality politics obscured his political views. The election in London was grimly fought, with all the big guns brought out in Johnson's support, and the swing away from Labour (Livingstone) was far smaller than elsewhere in the country, but Johnson won.

The first days after the shock of the election result were surprising, with the two candidates expressing mutual respect, even appreciation. Johnson spoke of building on Livingstone's achievements. It was a curious peace. Then, on May 25, Johnson made the first move that distanced himself decisively from the politics of his predecessor: he cancelled an agreement that Livingstone had established between London and Caracas, Venezuela.

One element of this agreement had been a simple swap. On the one side, London provided Caracas with technical advice and expertise on a range of urban issues, including transport planning, waste disposal, and environmental matters. A London office had been set up in Caracas for this purpose. On the other side, Caracas, through PdVSA, the Venezuelan state-owned oil company, provided London with oil at a reduced price. The oil was used specifically for public transport (buses) and made possible the reduction of fares by half for the poorest people in the city. (This was enabled through recalibrating people's electronic payment cards with the presentation of social security documents.) The aim was to use a transformed relation of international trade to effect redistribution within each city. In his response to the cancellation, Livingstone said,

> The basic principle of the London-Caracas agreement was simple, reasonable and indeed a rather textbook illustration of relative advantage in foreign trade. Each side provided the other with that in which they are rich, and which for them is therefore relatively cheap—oil, on one hand, and the expertise in managing a modern advanced city on the other—in return for something which was scarce, and therefore relatively expensive, for the other side.
>
> The benefits to the poorest people in London were evident—over 130,000 have benefited to date from half-price bus travel.
>
> The benefits to Venezuela were equally great. The accumulated expertise acquired by long-developed cities and companies is one of their most valuable assets. For Venezuela to develop this purely internally would take a very long time and be extremely expensive, while to purchase it from international consulting companies would cost many times that paid to London. (Livingstone 2008)

The cancellation of the agreement in May 2008 came just as rising oil prices meant that fuel poverty was dominating the media headlines. The new mayor chose that moment to remove some £15 million of fuel subsidy from the poorest among his electorate.

Yet much of the media supported the cancellation. Most termed the deal with Caracas, known as "oil-for-brooms," "controversial." The *Guardian,* a newspaper not viewed as on the right of the political spectrum, opined in an editorial that "Mr. Johnson did the right thing" (*Guardian* 2008, 28). The new mayor produced several rationales for the cancellation. He had long characterized the agreement as "completely Caracas" (completely crackers)—a play on words that added nothing to the political debate. He also, though, talked of saving £67,000 a year by closing the Caracas office and raised some more thoughtful political reservations:

> I think many Londoners felt uncomfortable about the bus operation of one of the world's financial powerhouses being funded by the people of a country where many people live in extreme poverty. I simply think there are better ways of benefiting Londoners and better ways of benefiting Venezuelans. (Mayor of London 2008)

Behind that, though, lay a deeper antipathy: in an interview recorded a hundred days into his administration, Johnson "refers to Venezuela's Hugo Chávez as a 'foreign dictator'" (Aitkenhead 2008, 27). Indeed, during the same period, the new mayor established an alternative foreign relationship—with New York and its mayor Michael Bloomberg. As a BBC News commentator reflected, "In the battle of ideologies, Mr. Johnson's Conservative administration would argue that New York . . . is a more suitable partner than Caracas" (Josephs 2008). Livingstone argued that "Boris Johnson's cancellation of London's oil agreement with Venezuela is a piece of rightwing dogmatism that is equally costly to the people of London and Caracas" (Livingstone 2008).

The postelection truce was over. It had ended, precisely, over a question of the politics of the city's relations with the wider world, over the politics of place beyond place (see chapter 5). I want to explore this agreement, and the politics around it, to examine the themes of this book.

Introductory Themes

A central theme of this book revolves around the relationship between *relationality* and *territoriality*. In the simplest terms, this might be read as the contrast between a focus on connections (and lack of connections), on the one hand, and a focus on places, on the other. The argument of this chapter is that, while this distinction might point to a difference in the starting point of analysis, or indicate a contrast in the central object of concern (relations or places), from the point of view of a relational approach, there is absolutely no conflict between them. The politics of the agreement between Caracas and London was a politics of *relations* (the nature of the "trade" relations between the two cities—there was also more to the agreement, as we shall see later), but this relationality was constituted through an explicitly *territorial* politics—a politics of place. From a relational point of view, moreover, the very identities of places (territories) are relationally constructed. That is to say, places are what they are in part precisely as a result of their history of and present participation in relations with elsewhere. London's identity, its very character as a place, is clearly a product of its long engagement with the wider world, as capital city of a nation-state, as trading center, as reigning city of worldwide empire, as crucial coordinator of financial globalization, as a focus of international migration that makes it today the most multicultural city on the planet. And so on. This is an extreme example, a particularly clear case, but the same is true, in principle, of all places. They do not come into being in isolation. This was an essential argument behind the idea of "a global sense of place" (Massey 1991). Territories are constituted and are to be conceptualized, relationally. Thus, interdependence and identity, difference and connectedness, uneven development and the character of place, are in each pairing two sides of the same coin. They exist in constant tension with each other, each contributing to the formation, and the explanation, of the other (Massey 1993).

This, then, is an argument from a relational perspective. There are extreme versions that argue that, in a global world, all is flow and connectedness and that no such coherences as "place," or territories of any sort, are either conceptually coherent or, empirically, viable for much longer. We are witness to the inevitable dissolution of place. A political version argues that, precisely because it territorializes what in principle is a global humanity, the notion of place at any scale (including the nation-state) is, in its divisiveness and in its generation of loyalties and thereby conflicts, necessarily politically reactionary (Hardt and Negri 2000). However, from a place-based perspective, it is possible, if only usually implicitly, to understand places or territories as closed entities that exist *prior to* their engagement with others. Such a conception is without a doubt opposed to a relational viewpoint. What is evident in the case of London–Caracas is that places do figure, that there is a politics of place that is the outcome of the contested negotiation of physical proximity, that this can be a politics of place that is explicitly relational with the beyond, and, finally, that such a politics need not be reactionary in a political sense.

Cities, then, as other places, are set within geographies of social relations (economic, cultural, political). Their character and this relationality are mutually constitutive. There are numerous lines of inquiry that may be pursued once given this recognition. The rubric for this book details two of them. The first is that the politics of cities will be concerned to manage the effect on the city of this global relationality. The second is that "urban policy actors" are involved in international networks through which urban policies are passed between cities and adopted or contested.

Both aspects of relationality will be evident in the case of London and Caracas. (Indeed, extending technical experience between the two cities could be seen as an example of the second.) However, cities can respond to their global relationality in other ways. In particular, the case explored in this chapter demonstrates how they can *use* their positioning within global power-geometries, how they can turn interdependence into an opportunity for change (Clarke et al. 2007). In other words, here the city governors go beyond managing the internal impact of the relations their cities are set within and try to change those relations (chapter 6). This, then, is a second, introductory, theme.

A third theme is that this approach to understanding the inter-dependence of places can be the basis, not only for analyzing that interdependence, and not only for producing a critique of its present form ("critical geography"), but also for the construction of alterna-tives. Both London and Caracas, in their official voices and to a large extent in their grassroots politics and quotidian lived relations, posi-tioned themselves in opposition to U.S. hegemony in general and to the economics of the Washington Consensus in particular (Massey 2007; Cariola and Lacabana 2005a; Massey and Livingstone 2007). Caracas has been written of as both global and counterhegemonic (Lacabana 2008). Lacabana writes of the knowledge circuits (see chapter 4), here of the political Left, that converge in the city, and of the focus there of networks of social movements (see chapter 5). The agreement between the two cities was one small part of wider attempts to construct alterna-tive relations that proposed a counterhegemonic solidarity. In doing this, both place and relationality were crucial.

Two Trajectories

London has "reinvented itself," as the London Plan puts it, since the 1980s. In that decade of the abolition of the Greater London Council (GLC) by Margaret Thatcher, it was a city that figured in the popular and political imagination, primarily in terms of the collapse of the old docks, the decline of manufacturing industry, and the classic problems, here writ large, of inner urban areas. (Of course, it figured as a place of political contest.) Since then, and partly because of the political victory of the right in that decade, London has established itself as a preeminent world city, a crucial node in the organization, diffusion, and maintenance of *neoliberal financial globalization*. This designation immediately indi-cates that what "won" in the political contests of the 1980s was, in fact, a small part of the metropolitan area: the financial City (capital C) and the politico-economic philosophy to which it was committed. That City was, in the beginning, a prime mover in inventing this new world of "neoliberal globalization."[1]

It remains as the financial center of this new imperialism. All kinds of relations run out from London to the rest of the planet—investment, trade, security, insurance, speculation—some beneficial, many pernicious, that are at the heart of the production of the current form of globalization.

Yet, as we have seen, London is a politically "progressive" city. Indeed, there are many ways of being a world city, and London has become one in terms of its cultural mixity. When Livingstone, then mayor, spoke in July 2005 after the bombing of public transport, it was this that he evoked (GLA Press Release, July 8, 2005; Massey 2007). Surveys show this is how Londoners think of London and what they like about it (which is especially the case with ethnic-minority Londoners). Gilroy (2004) has written of its "demotic cosmopolitanism." After the bombing, across the political spectrum, the assertion was "London stands for this."

There is, then, a contradictory positioning. This is a city at the very center of the reassertion of marketization, profit, and privatization, which yet imagines itself (and not incorrectly) as open, as hospitable, indeed, in a certain sense, as generous to the outside world. Further, there was until May 2008, a mayor who was explicitly positioned against the Washington Consensus, yet in whose parish the infrastructure of global neoliberalism flourished and flaunted as never before. This is a city indeed whose very multiethnicity derives in part from people coming here, not only because of the freedom of the place (which is real), but because their lives elsewhere have been destroyed by forces some of which originate in London. Over the past thirty years, this has been London's trajectory.

Meanwhile, over the same thirty years, across the Atlantic to the south, other trajectories were being woven. On the northern coast of Venezuela, the red mountains plunge straight down into the sea, and the plane lands on a narrow coastal strip with only dunes hiding the Caribbean. Through the mountains lies another capital city—Caracas—of a country that extends to the Orinoco and the northern headwaters of the Amazon. It seems a world away from London, but here too one key sector (like finance in London) is the link to the global economy. Here, it is oil.

Oil had been nationalized in Venezuela in 1976. From the 1980s, elements within the elites of Caracas faced, as elsewhere, by the crumbling old settlement, had become enthused by the new neoliberalism. The economy was to be opened up to the world, and opening up oil was key to this. The well-known mantras were pronounced of the need to cut back the state, to privatize, to rewelcome the private oil majors, to reinsert Caracas into the hierarchy of world cities, for this was now the aim of almost every city plan around the world.[2]

As it had been in London, this process was contested in Caracas. In 1989, simmering heterogeneous anger burst out in the Caracazo with two days of protest and violence on the streets in which many hundreds were killed. In 1992, a coup attempt ended in failure. Then, in December 1998, the old order was more thoroughly challenged when Hugo Chávez was elected president. Since then, he has survived a coup attempt (2002), a petrol lockout and strike (2002–3), and a recall referendum (2004), all of which the U.S. government, at least tacitly, supported. In 2006, Chávez was reelected to a second term as president.[3] Over these years, his mission has evolved as the Bolivarian project for a socialism of the twenty-first century.

Cities are widely recognized as crucial to the neoliberal world order. They are growing rapidly. The competition between them is both product and support of the neoliberal agenda, and concentrated in them are the social strata and the institutional and cultural infrastructure, which is key. That global order in turn affects cities. Caracas had already been a very unequal city, but it had become more unequal with the neoliberal hegemony (the Caracazo was one response). Lacabana and Cariola, in a number of research projects, have documented the changes in the city, before and since the election of Chávez (Lacabana and Cariola 2001, 2006; Cariola and Lacabana 2005a, 2005b). New structures of grassroots, participatory democracy have been established, gradually enabling the city's poor to find a voice. And the income from oil is being used within the city (and elsewhere in the country) to fund a range of projects—in housing, health, popular education, and production cooperatives. Although still incipient, and hotly contested, some changes are evident. Lacabana and Cariola record, for instance, a breaking out of the poor from their previous enclosedness in their own parts of the city, through the development of collective space, an expansion of institutional space, and a spatial opening up of daily life.

This, then, is more than managing the effects of the city's positioning within the power-geometries of globalization. It is an attempt, through using the income from the key link into the global economy (oil), radically to change the internal contours of the urban area.

Place beyond Place

However, this Bolivarian project has more than local ambitions. It is explicitly working for change on a wider geographical canvas. This it does in a whole range of ways and through a constellation of different relational geographies. The Bolivarian Alternative for Latin America (ALBA), an incipient politico-economic alliance of states within Latin America, is one central element, with the aim of establishing an alliance that is not of the free-market competitive nature of its attempted predecessors (Katz 2007). It is also an attempt to assert greater autonomy from the United States. So too are Telesur (an alternative to CNN) and the Banco del Sur (an alternative to the International Monetary Fund and the World Bank). Teachers and medical teams fan out across Latin America. The World Social Forum (2006) and the World Women's Congress (2007) have been held in Caracas. A range of multilateral and bilateral agreements and exchanges adds to these attempts to shift just a little of the power-geometries of the hegemonic form of globalization. And central to these is oil. Through PetroCaribe, most of the countries of Central America and the Caribbean receive oil at reduced prices; there is an exchange of oil for doctors with Cuba; through CITGO (a Venezuelan-owned petroleum distribution company) and through the Citizens Energy Corporation and a range of Inter Tribal Councils, free and reduced-price oil has for some time been distributed to the United States. The aim is to work against the Washington Consensus and its successors and against that "neoliberalism" at whose birth and in whose development and current propagation London has been so important.[4]

Within Venezuela, Caracas is central to all of this. As part of this international project, there is an explicit repositioning of Venezuela, and of Caracas in particular, within the power-geometries of globalization. Cariola and Lacabana (2005a) write that what is at issue is not just managing the effects of globalization, or withdrawing from global involvement, but reorganizing the nature of international relations. As Cariola and Lacabana write, the aim of turning Caracas into a classic global city has been replaced by an attempt at using the forms of the city's interdependence to challenge existing terms of globalization.

It is now widely accepted that local places, such as cities, are formed in part as the products of wider relations of globalization. This is the characteristic implicit imagination of the relation between local place and globalization. One problem with this conceptualization is that it can easily slip into an interpretation of local places as *victims* of globalization. Globalization (or, more generally, geographically more extensive relations) is imagined as an external force that arrives to affect the local place. The direction of relationality, in other words, is conceptualized as being one way, from the outside in. (And on occasion, this may be an inaccurate picture.) In general, however, those relations with the wider world are two way. Globalization is not always an external force, arriving from somewhere else. Logically, anyway, this is impossible. For globalization is made in places. As there is the global production of the local, there is also the local production of the global. In much counterglobalization discourse, the tendency is to exonerate the local; the local is positioned as recipient rather than as active agent. Yet in most places—certainly most cities—a counterglobalization strategy should involve not simply defending the local against, or even managing the local's position within, the global, but also *challenging* the city's place within, and contribution to, the wider global structures—challenging and changing the nature of the local insertion into the global, and thus the effects of that insertion. This effort to change the nature of the city's insertion into global structures, and thus change its effects and its influence, is precisely what is going on in Caracas. One element in this strategy—one small element—has been the agreement with London.

Meanwhile, while this political contest, imagination, and invention had been going on in Caracas, the new economic world order also had effects being contested in London. Within London, inequality has increased, especially as the result of the rise of a stratum of the superrich; the presence of this stratum, through the price system, works to make the poor poorer; the same price effects make the social reproduction of the city more difficult; and the dominance of the finance constellation makes even more precarious the survival and growth of other sectors.[5] An immediate concern is that two modes of London's reputation as a "world city" (its multicultural mixity and its global financial supremacy) will come into conflict. That is to say, a serious danger exists that the inequalities produced by the financial world city will undermine the (relatively) good relations that currently hold together its multicultural character.

It could be argued that any left-wing administration in London is caught in a trap. Finance, unlike the oil sector in Venezuela, is not in public hands. In other words, the social relations of the city's chief sectoral mode of insertion into the global capitalist economy are quite different. Moreover, London's city government faces a national government that is unswervingly committed to the financial Square Mile, and all it stands for, and to the strata who benefit from its dominance (and its deregulation). So London's options, the room for maneuvering its city government, are more constrained. In that context, the policy directions under the Livingstone mayoralty were complex. First, huge attention was paid to the growing inequality in the city. This took the form of thorough research and documentation and genuine commitment of policies. Policy decisions in a range of areas, including transport, were designed to reduce economic and social inequality. A "living wage," above the national minimum wage, was agreed to and implemented. There was a complex program of contract compliance. All this, however, was designed to address the *effects* of the new economic order. It could be interpreted under this book's rubric of "managing global relationality." However, the irony in this case was that much of this global relationality has its sources within the very same city.

Second, however, the mayoralty under Livingstone also *supported* the financial city of London and its role in the world. The plan for the metropolis indeed presented globalization as though it came from somewhere else (as though the agency for its production could not be found within London). The plan also hailed the role of the financial global city simply as an achievement, as something the city should be proud of. It did not inquire as to the effects of that role either within the country as a whole or on the wider world.

Yet the mayor was also cannier than that. He was quite clear in accepting the new hegemony as an implacable fact.[6] Indeed, he saw no way of challenging that hegemony within the city. As he argued, the financial constellation provided the most important employment base of the urban area. Moreover, he argued, any attack on these sectors, and on their role in the wider world, might only provoke them to leave. (These arguments can be responded to. First, as Livingstone said in the same discussion, there is nowhere else in Europe the financial City can move to. There are also a range of advantages to being in London. Second, this position ignores the point that, rather than being only the golden goose, as it so often claims, the financial City, in its structural dominance, has long been and is even more so now, *damaging*, socially and economically, even within the urban area.) However, Livingstone's argument was that it was not possible to use the financial sector in London to challenge international arrangements in the way that Caracas uses oil. He agreed that "there's this real contradiction—at the heart of an administration that's making a real case for international redistribution of wealth and power, I'm reliant on what is probably the world's biggest concentration of global corporate power outside New York" (Massey and Livingstone 2007, 20). Given this, he says, "you've got to build global structures of progressives and Labour and Greens to tackle it" (21). In other words, building alternative international relations challenges the structures of hegemonic globalization: "London's role is to do everything possible to encourage links—between what's emerging in China and India, the progressive forces in the West, progressive forces in Latin America and so on. We just play a role in encouraging, helping develop all that" (23). And one link he established to do this was the agreement with Caracas.

The Caracas–London Agreement

The Caracas–London agreement thus emerged out of a number of forces. In general, it arose in a context in which the mayoralties of both cities were explicitly politically positioned against both the type of global economic order embodied in the Washington Consensus and its successors and the hegemony of the United States. From both cities came the desire to challenge hegemonic economic and political power-geometries. For both cities, indeed, this particular agreement was one element in a wider strategy. The idea was first mooted when delegations from the two cities met and talked at the World Social Forum. It was rooted in and grew out of the politics of place in both cities. One element of the agreement, not yet mentioned, was that a program of cultural exchange would be established, alongside the oil-for-brooms deal, that would enable the people of the two cities to get to know one another—and their politics—better. The media response to the cancellation of the agreement (see previous text) demonstrated the depth and generality of the hostility among the establishment in the United Kingdom toward the political changes under way in Venezuela. This aspect of the agreement was only beginning to develop when it was cancelled, but it constituted a small attempt at building solidarity as a process rather than simply a declaration.

The exchange of oil for technical experience, however, emerged from an understanding of the *relational* positioning of the two cities. For Caracas, and Venezuela, oil is the main form of economic interdependence with the global economy, and the decision was made to use that interdependence to change wider power-geometries. In London, the mayor had deemed it impossible to use its own major sectoral link to the global economy (finance) in the same way. The route therefore taken was to build alliances that might present a challenge internationally.

The agreement was an attempt explicitly not to engage in the market relations of the current form of globalization. It was an attempt to bypass them and thereby show that they could be challenged; it was an attempt at equal exchange.[7] It was also a clear challenge to that neoliberal mantra that places in general and cities in particular must compete against one another. A stance that asserts that, including in economic exchange relations, it is possible to cooperate. Moreover, it demonstrated that this is possible even between places in different positions within the wider power-geometries of globalization. It was a demonstration, through the reconstruction of one set of relations, of an alternative globalization. At the same time, it was also a form of symbolic politics, an exemplary project: it stood for a wider proposition. The intention was to puncture the "normality" of our "normal" assumptions.

Moreover, and to pick up again the argument about relationality and territoriality, given time this rearticulation of relations would or could have fed back into how the identity of each city (territory) was constituted. Here, place (territoriality) and relationality are two sides of the same coin, mutually constitutive and always in play together. Mutual knowledge could have been one element in this, together with relational empowerment. The accord was also a jolt to the self-perception of both cities in that it asserted, through the changed exchange relations, their equality. This may seem trivial, but this was an accord between a place widely seen as one of *the* global cities and a place, far smaller and historically colonized, within the global South. When one prominent political commentator in London, criticizing the deal, exclaimed that "it makes us feel like a Third World country," it was possible to detect that jolt to the dominant, imperial, geographical imagination that had been provided and the implicit questioning of the identity of places that went along with that. Indeed, the disagreement and debate that took place fiercely in Caracas also was a way of opening up the identity of place to wider democratic engagement by posing the implicit question: what does this place stand for? (Massey 2007).

Final Reflections

That the accord between London and Caracas was cut short after only a few years means, of course, that we cannot know the degree to which its potential effects might have been realized. More positively, its cutting short clarifies the processual nature of solidarity (Featherstone 2008)— both the agreement and the nature of the relations it embodied, and the debates it had sparked within both places, were continuing to develop. The cutting short also means, therefore, that (from my perspective) many aspects of the accord remained misrecognized and misunderstood. This chapter aimed to spell out my interpretation of its geographies and of its politics, both realized and potential. In the context of the arguments of this book, however, the agreement was a working example of the constant interplay between territoriality and relationality and the ongoing trajectories of mutual modification. Imagining global space as either pre-established closed territorialities or a smooth space of flows is to ignore the grounded and interconnected reality of real politics. The agreement was an example too of territorially defined political agencies attempting not only to manage the effects of the wider world on the internal workings of those places, nor only to engage in the circuits through which policies themselves circulate between places (though it related to both of these things), but also to go beyond both of these and intervene in the nature of global power-geometries. Finally, the political intent of the intervention was to be, explicitly, counterhegemonic.

Notes

1. This contested history of London and, indeed, many of the arguments here, can be read in more detail in *World City* (Massey 2007). That work also considers the complexities and ambiguities surrounding the term *neoliberalism*, for which there is insufficient space in this chapter.

2. This is, necessarily, an extremely bald account. For a fuller and more nuanced account, see Massey (forthcoming). See also Gott (2005), Wilpert (2007), and the references here to the work of Miguel Lacabana and Cecilia Cariola.

3. In December 2007, he lost a referendum on changes to the constitution, but his changes were approved in a February 2009 referendum.

4. Wilpert (2007) attempts a systematic political assessment of the various strands of post-1998 politics in Venezuela.

5. For documentation of all of this, see Massey (2007).

6. The arguments that follow can be found in an interview with Livingstone (Massey and Livingstone 2007).

7. This element of the agreement was not argued strongly enough in the political arena. It was partly for that reason, I believe, that the purpose of the agreement was misunderstood.

References

Aitkenhead, D. 2008. "Capital Charmer." *Guardian*, August 9.

Cariola, C., and M. Lacabana. 2005a. "Globalización y Metropolización: Tensiones, Transiciones y Cambios." Mimeo. Caracas: CENDES-UCV.

———. 2005b. "Los Bordes de la Esperanza: Nuevas Formas de Participación Popular y Gobiernos Locales en la Periferia de Caracas." *Revista Venezolana de Economía y Ciencias Sociales* 11, no. 1: 21–41.

Clarke, N., C. Barnett, P. Cloke, and A. Malpass. 2007. "Gobalising the Consumer: Doing Politics in an Ethical Register." *Political Geography* 26:231–49.

Featherstone, D. 2008. *Resistance, Space, and Political Identities: The Making of Counter-Global Networks*. Chichester: Wiley-Blackwell.

Gilroy, P. 2004. *After Empire: Melancholia or Convivial Culture?* London: Routledge.

Gott, R. 2005. *Hugo Chávez and the Bolivarian Revolution*. London: Verso.

Guardian. 2008. "From Caracas to Crossrail." May 27, 28.

Hardt, M., and A. Negri. 2000. *Empire.* Cambridge, Mass.: Harvard University Press.

Josephs, J. 2008. "Living Up to the Manifesto Pledge." BBC News. http://news .bbc.co.uk/go/pr/fr/-/1/hi/england/london/7419533.stm.

Katz, C. 2007. *El Rediseño de América Latina: ALCA, Mercosur y ALBA.* Caracas: El Perro y la Rana.

Lacabana, M. 2008. "Caracas: Global y Contrahegemónica." Paper presented to X Seminario Internacional RII, Santiago de Querétaro, Mexico, May 20–23.

Lacabana, M., and C. Cariola. 2001. "La Metrópoli Fragmentada: Caracas Entre la Pobreza y la Globalización." *EURE: Revista Latinoamericana de Estudios Urbano Regionales* 27, no. 80: unpaginated.

———. 2006. "The Processes Underlying Caracas as a Globalizing City." In *Relocating Cities: From the Center to the Margins,* edited by M. Amel, K. Archer, and M. Bosman. Lanham, M.D.: Rowman and Littlefield.

Livingstone, K. 2008. "A Piece of Mindless Vandalism." Comment is Free. *Guardian.* http://www.guardian.co.uk/commentisfree/2008/jun/02/apieceofmindlessvandalism.

Massey, D. 1994. "A Global Sense of Place." In *Space, Place, and Gender,* 146–56. Oxford: Polity.

———. 1993. "Power-Geometries and a Progressive Sense of Place." In *Mapping the Futures: Local Cultures, Global Change,* edited by J. Bird, B. Curtis, T. Putnam, G. Robertson, and L. Tickner, 59–69. London: Routledge.

———. 2007. *World City.* Oxford: Polity.

———. *Voices of Places.* Oxford: Oxford University Press, forthcoming.

Massey, D., and K. Livingstone. 2007. "The World We're In: Interview with Ken Livingstone." *Soundings: A Journal of Politics and Culture* 36:11–25.

Mayor of London. 2008. "Mayor Brings Controversial Venezuela Deal to an End." News release. http://www.london.gov.uk/view_press_release.jsp?release id=17053.

Wilpert, G. 2007. *Changing Venezuela by Taking Power: The History and Policies of the Chávez Government.* London: Verso.

CHAPTER TWO

The Spaces of Circulating Knowledge

City Strategies and Global Urban Governmentality

Jennifer Robinson

At some stage, most cities around the world have been or will be the subject of a long-term citywide strategic policy exercise. Often, they are undertaken at moments of political change or local crisis. Usually, they result in a focused statement visioning a city's future as well as more substantial programmatic policy documents that lay out specific plans for some combination of improving the quality of life for residents, promoting economic growth in the city, and ensuring long-term sustainability. These city strategies are more or less ubiquitous: larger and smaller; wealthier and poorer cities in all regions of the globe can and do undertake such exercises. They are promoted through a range of different technical and professional rubrics—spatial planning, international urban development policy, environmental sustainability, urban economic policy, and city marketing. Rather than one uniform policy technology, I am referring to a family of different strategic policy initiatives focused on the city scale, with divergent regional incidences and origins, but much overlap of policy emphasis and personnel. They assume similar formats—usually strong technical input, with consultative processes with stakeholders and publications targeted at shaping opinions of key decision makers and the public. They share remarkably similar analyses, conclusions, and policy ambitions. They are actively promoted as tools for urban development and planning by networks of city managers, by independent consultants, and by international development organizations (see chapters 3 and 4 for similar efforts).

Most studies to date consider these initiatives within relatively limited national, regional, or developmental frameworks—mapping, to some extent, the origins and reach of different city strategy methodologies. A robust literature on boosterism, place competition, and city governance exists within the U.S. context (classically, Logan and Molotch 1987; Cox and Mair 1988), and some studies have been undertaken in a range of other cities around the world, often assessing these in the light of the U.S.-based literature (Machimura 1998; Jessop and Sum 2000). The experiences of cities in the United States and the European Union (EU) have been explicitly compared by several scholars (e.g., Vicari and Molotch 1990; Harding 1994). Cities within a developing country context that have undertaken these exercises are more likely to be considered within the context of debates on international development policy where City Development Strategies (CDS) have come to form one component (with Slums Upgrade) of the two-pronged strategy of the Cities Alliance initiative of the World Bank and United Nations Centre for Human Settlements (UNCHS) (Harris 1995; Stren 2001; GHK Consultants 2002).

In contrast to this currently fragmented literature, I suggest that these relatively ubiquitous long-term strategic urban policy exercises—city strategies—would benefit from an assessment through a properly global and transnational analytical lens. This chapter does some conceptual ground clearing to consider how one can approach the study of urban phenomena, which are global in terms of the reach of the networks of knowledge production associated with their preparation; ubiquitous in terms of their relevance and take up; regionally differentiated in terms of the tracks of policy influence that have shaped them; and often profoundly localized in terms of their politics and content. Importantly for our assessment of them, they are also characterized by a range of power relations associated with their adoption and execution, from imposition through hierarchical institutional expectations to enthusiastic local or popular appropriation with strong participatory methodologies (see chapters 4 and 5).

Approaching the phenomenon of city strategies therefore raises substantial theoretical challenges for contemporary urban studies. These include how to characterize the complex spatialities of a globalized policy instrument; how to assess the power relations involved in adopting and disseminating ideas and practices of city strategy formation; and how to understand the political meaning of the resultant policy documents. More than this, these policy exercises also require an interrogation of the suitability of extant urban comparative methods for pursuing the study of cities "at a world scale" (Connell 2007; Robinson 2008). In the field of urban studies where wealthier and poorer cities seldom have been considered alongside one another, city strategies provoke an exciting engagement with urban politics and processes across a much wider range of cities (see chapter 1). They challenge urban scholars to address the historic analytical divisions between cities assumed to be rendered incommensurable by their developmental status (Robinson 2006). The most substantial wager of this postcolonial critique is that drawing the world of cities closer together, in this case through attending to city strategies within a transnational perspective, could stimulate a reevaluation of some core topics in urban studies, such as (in this case) circulations, global governance, and local politics.

After a brief introduction to the phenomenon of city strategies, this chapter turns to an exploration of the spatialities of circulations. These are exposed by the ubiquitous phenomenon of city strategies and the power relations of global governance that frame their adoption, especially the scope for local autonomy in this realm of globalized policy formulation. Overall, by focusing on city strategies, the chapter offers an analytical approach to a substantially transnationalized world of urban policy formation encompassing both wealthier and poorer cities and advocates for the importance of a more properly postcolonial urban studies.

City Strategies

City strategies are undertaken by cities for many reasons. Often, moments of political or economic transition encourage city managers to reevaluate policy directions and opportunities. Many national governments require routine periodic revisiting of strategic policy directions for cities. And, in times of crisis, city governments might turn to strategic visioning to imagine appropriate responses (McCann 2001; Lipietz 2008). While comprehensive city strategies undertaken for these different reasons have many features in common, such as a desire for inclusivity or popular participation in decision making, the use of expert technical advice, or a similar range of policy ambitions and regulatory tools, they emanate from a number of different sources of inspiration. Spatial "structure" planning practices, for example, have acquired a more strategic direction, often making comprehensive spatial plans more akin to city strategies (Healey 2007). Planning for sustainable urban development, in taking a comprehensive citywide view of sustainability, has similarly produced policy exercises that are in effect city strategies (Myers 2005). Cities are often supported or encouraged by private sector consultants to embark on major strategic reviews, and these consultants might offer their own formulas for conducting such a city strategy (Robinson and Boldogh 1994; McCann 2001). Some cities have been encouraged to prepare City

Development Strategies by international development agencies (GHK Consultants 2002; Cities Alliance 2006a).

City Development Strategies are required by some national governments. For example, in the United Kingdom (UK), the Greater London Authority is required by central government to prepare a series of eight plans, including economic development, transport, and housing, with the spatial development strategy, articulating an overall strategic London Plan (Travers 2004). Across the UK, Community Strategies (CS) generated through participatory processes have provided a broad framework for local government policies since 2000 (Raco et al. 2006). In South Africa, integrated development plans for five-year planning cycles in metropolitan areas are now expected to be linked to longer-term City Development Strategies, making the conduct of these strategic exercises a routine part of city government (Harrison 2006). In his review of the South African medium-term Integrated Development Plan legislation, Harrison (2006) specifically comments on the similarities between the UK community strategies, and the Integrated Development Plans/City Development Strategies conducted in South Africa. In a wider European context, strategic planning approaches have been consolidated and promoted, resulting in expectations of city visioning and strategic thinking by municipalities in some national contexts and, notably, among new members of the EU (Healey 2007).

An important vehicle for articulating the technique of City Development Strategies has been the Cities Alliance, a joint initiative of UNCHS and the World Bank, which has popularized CDSs in most regions of the world, notably Southeast Asia (Asian Development Bank 2004). This grouping has two core programs, Slum Upgrading and City Development Strategies, both aimed at cities in poorer country contexts. They have prepared a substantial document outlining how to conduct a CDS and explaining why they are valuable exercises (Cities Alliance 2006a). Funds are available to provide technical support for undertaking a CDS, and securing financing for long-term development projects identified through the CDS is a strong ambition of this initiative. The CDS approach is intended to be inclusive and participatory, invoking strong local leadership in close association with other levels of government and various potential partners from the private sector, communities, and nongovernmental organizations; international donors; or other government agencies. There is a strong developmental agenda in relation to service delivery, alongside an assumption that economic growth is essential for meeting the substantial social and environmental challenges in poorer cities.

Whatever the sources of inspiration for conducting a strategic review, city strategies have many concerns and discourses in common. Cities are understood to need to frame their futures within the ambit of the global economy—to find their own place in a rapidly changing and usually challenging economic world. City strategies tend to see the need to adopt a balanced approach to development; economic success is seldom identified as the only ambition. Quality of life and sustainability commonly accompany economic growth as the core aims of a city strategy (chapter 4). A preliminary review of city strategies provokes a powerful sense of repetition and replicability in themes and approaches across a wide range of different urban contexts.

The technique of doing a city strategy circulates internationally, although different histories and wider tracks of influence on specific cities cause certain policy ideas rather than others to become prominent within the framework of these exercises in particular places. The common priorities articulated across numerous city strategies—economic growth, environmental sustainability, quality of life, and fiscal stability (see World Bank [2000] for an articulation of these common priorities)—are potentially brought into relationship with one another differently. They might be balanced through particular locally defined trade-offs, or policy choices, or prioritized as a result of some political conflict or strategic policy direction, or simply assembled as concurrent policy ambitions. In terms of urban theory, this important feature of city strategies deserves further attention. In taking a citywide view, city strategies can potentially counteract theoretical segmentations of much contemporary urban studies, where economic processes, environmental challenges, cultural politics, and social change are often considered in isolation (Robinson 2006, chap. 5). Strategic visioning forces citizens, consultants, and urban managers to think across the city as a whole, bringing elements often analyzed in isolation into relationship with one another. What this also means is that while they seem to have many common features, and are closely tied in to international policy circuits, city strategies emerge from and are highly responsive to local specificity and thus could potentially generate appropriate and positive development plans for individual urban areas (Parnell and Robinson 2006; Healey 2007).

The policy of supporting City Development Strategies to promote urban policy formation in poorer cities (Cities Alliance 2006a) has been subject to some criticism (Stren 2001; Pieterse 2008). Concerns expressed in response to the formal policy documents include their apparent complicity with the broadly neoliberal development agendas of international agencies and their overly technicist approach, which Pieterse (2008) suggests sidelines the tough politics of negotiating city futures across the divisions and conflicts that characterize many cities, especially in resource-poor contexts. They are also vulnerable to political change—major policy visions can often be overturned by new political leaders (Stren 2001; GHK Consultants 2002; Cities Alliance 2006b). Substantial amounts of scarce resources have been directed at these comprehensive city strategies—evaluating their effectiveness and potential is important. However, I share with Patsy Healey a sense of hopefulness about the political potential of these exercises (Parnell and Robinson 2006; Healey 2007). As she notes in relation to her EU case studies of strategic planning, there is "the possibility of a strategy-making that has the capacity to pursue a rich and diverse conception of an urban area which holds concerns about distributive justice, environmental well-being and economic vitality in critical conjunction, rather than a narrow focus around the objectives of a few actors and social groups" (Healey 2007, 32).

In poorer country contexts, the significant challenges of basic service delivery and political inclusion layer in an urgency to her hopefulness that the Cities Alliance (2006b) and other international agencies are eager to convey through their attempts to popularize City Development Strategies (UN 2006). Contributing to efforts to improve governance and to develop and implement robust and effective policies to address the overwhelming needs of urban dwellers in rapidly growing poor cities has to be a priority for urban scholars and practitioners around the world (Davis 2006). City strategies have been set in motion in the global arena of urban policy formation, sometimes in the hopes of contributing to meeting these challenges, indicating that assessment of their potential and effectiveness is important.

This requires that we think across the range of different contexts in which city strategies have been invented and implemented—that is, assess them as a global urban policy technique, whose circulations and meanings extend beyond one region or category of cities. Therefore, they are neither simply an evolution of a relatively progressive Western planning tradition (as Healey 2007 sees them) nor a disciplinary technique developed by international development agencies for control of poorer cities, as Pieterse (2008) argues. This suggests that we should approach city strategies with some care in assessing the power relations involved in their uptake and execution. Healey's (2007) extended analysis of comprehensive strategic planning in Europe draws on a sophisticated understanding of power, including careful institutional analysis, attention to the subtle power of discourses, and awareness of the overt political contestation in many cities over urban policies. I sense that to do justice to the assessment of city strategies we need to deploy just such a subtle analysis of power.

For the purposes of this chapter, then, city strategies provoke some important directions of inquiry. First, as a genre of urban policy, they stretch comfortably across wealthier and poorer contexts. Thus, they demand that we extend our analysis to incorporate a wider world of cities than that which conventionally underpins urban theorizing. In this vein, city strategies inhabit a world of global urban policy that has as yet drawn little attention from urban studies (although see Cochrane [2006, chap. 7]; McCann [2011]). Global urban policy, here, indicates the potential for policy ideas to circulate in all sorts of different directions across the globe, stretching across a wide range of different contexts; the circuits, or switches, of policy transfer are multiple, even indeterminate. It rests on an observation that there are agents of policy circulation whose ambition is substantially transnational: whether these are International Financial Institutions (IFIs) aiming to spread good practice to as many poorer countries as possible, private sector consultants with a strong interest in finding new clients, or managers of cities eager to see their achievements broadcast on a world stage. Here, global does not signify a single planetary circuit of knowledge but, rather, overlapping and multiple transnational tracks of engagement, influence, and information.

Many accounts of urban policies in wealthier countries offer national-centric narratives of policy developments, with only a few specific engagements with the wider sources of policy ideas and practice, mostly around an EU–U.S. policy transfer nexus (see, for example, Travers 2004; Ward 2006). Often, questions about the international circulations of policy are not explored in urban contexts in wealthier countries—an assumption is made that policy innovations have a national trajectory and provenance (in the UK see, for example, Raco's [2007] otherwise excellent study). In poorer country contexts, the opposite assumption is usually made, and one of the first questions asked of a policy initiative is, where did it come from?—although, often, policies are local translations or inventions. The following section of this chapter takes up establishing the grounds for more global assessments of urban policy, specifically considering how to conceptualize the spatiality of urban policy circulations appropriately across both wealthier and poorer country contexts.

The second compelling issue raised by city strategies concerns the form of power relations that they crystallize, especially, for our purposes, the power relations associated with their conception, circulation, and uptake. Once again, bringing conventionally divided approaches to wealthier and poorer cities into the conversation will be useful to appreciate these dynamics. If observers of wealthier cities seldom raise questions concerning the relative power of countries and cities in relation to the policies they adopt—and relative autonomy is more or less assumed for these entities—in poorer country contexts, the too easy assumption might be that international agencies, donors, and other powerful actors impose policy agendas on relatively weak local government agents. Building from accounts of global governmentality, the second section uses the example of city strategies to dispel this easy stereotyping of differential power relations, and to indicate how studies of wealthier and poorer cities could learn much from one another in terms of understanding the power relations of policy uptake and outcomes. The chapter closes with a preliminary assessment, based on the range of examples discussed, of the potential of city strategies to mobilize creative local responses to urban challenges and to frame productive, if sometimes strenuous, contestations over urban futures in particular localities.

Spaces of Circulation

How do ideas about city strategies and the urban development policies they embrace emerge in different contexts—either as repetition, as borrowing, or as new invention? Much is at stake in how we characterize the spatiality of urban policy transfer and learning. It is important to question understandings of policy exchange and innovation that are the inheritance of a deeply divided urban studies shaped by colonial and developmentalist assumptions. Conceptualizations of the power relations of learning (often assumed to be imposed by powerful Western or international development agents), deeply embedded assumptions about creativity and mimicry as the preserve of wealthier contexts, and presumptions about the trajectories of learning all need to be questioned. In this regard, the vocabulary we use can perpetuate certain assumptions about power relationships—"trajectories" of policy learning, for example—imply directionality and thus a sense of a distant origin and mimicry on the part of the receiving context and can also tend to imply a form of imposition. As an alternative conceptualization, "networks" of policy exchange might flatten out the horizon of policy transfer and evacuate the often uneven, if subtle and dynamic, power relations at work in policy formation and practice. As we explore a transnational—"global"—policy phenomenon like city strategies, then, we need to explore alternative vocabularies and conceptualizations of the spatialities at work in processes of globalization.

Few existing studies of urban policy learning examine their specific trajectories. Ward (2006) offers an excellent example, however. He carefully traces the lines of influence that precipitated the implementation of Business Improvement Districts (BIDs) in the UK, largely influenced by developments in the United States. He explores how a model of BIDs was produced in certain New York districts, shaped by an earlier Canadian initiative as well as by previous U.S. district management policies, and how this came to be appropriated—albeit in a somewhat different form—by UK policymakers. Ward wants to be "specific about where this version of 'the model' originated" (60). He tracks both a general interest among UK policymakers in the United States, and in New York in particular, and a very specific series of policy transfers, through central state motivations, circulations of documents, hosting conferences, popularizing exemplary case studies, and organizing personal visits and interactions, around the BIDs. So policy developments in the UK more generally (a neoliberalization of government) and a broad policy orientation to the United States dovetailed with specific attempts to promote a form of BIDs there. Interestingly, the historical origins of BIDs in Canada was forgotten in the flurry of links developed by a wide range of agents between the U.S. BIDs experiences and UK actors concerned with business involvement in town center management. Ward comments that, in the process of this (neoliberal) policy transfer, local government agents were remade in the image of neoliberalism—as "more enterprising, more entrepreneurial, more neoliberal" (71). In conclusion, though, he comments that "we actually still know very little about the details of how neoliberalism is made in different places, and the mechanisms through which 'local' policies, practices and programs are constructed as 'models'" (71). Perhaps, most important, he suggests that we have much to learn about how policies, or neoliberalisms, are created differently in different places where policies are appropriated or imposed in their name (see also chapters 3 and 4).

I suggest two ways to extend Ward's insightful exercise in tracking the routes of a specific episode of policy learning. The first, which I will return to in the following section of the chapter, picks up on his encouragement to explore how policies are made differently in different localities. I will draw on accounts of global governmentality to develop the idea that creative and differentiated policy outcomes can proliferate—even under conditions of neoliberalism. Second, drawing on the phenomenon of city strategies, I want to go beyond his metaphor of policy trajectories to suggest that the spatial dynamics of policy learning are considerably more complex than this practice of tracing trajectories implies. Specifically, we need to develop a spatial vocabulary more adequate to capturing the spatialities of circulation.

The case of city strategies is especially useful in illuminating the multiple and often overlapping circuits of urban policy that operate in a context of substantial policy globalization. The tactic of comprehensive city visioning has traveled across a number of different professional and discursive terrains—spatial planning; urban economic strategy; political boosterism; international, supranational, national, and intranational funding decisions; local environmental planning; and international development practice. The individuals, firms, organizations, and networks involved in City Development Strategies overlap considerably, and yet I sense that segmentation and differentiation continue to shape policy circuits: an individual consultant might work on a formal spatial plan, a bid for regional funding from the EU, and a City Development Strategy funded by the World Bank (G. Clark, personal communication, March 2008), while the policy trails of EU and international urban development practice may find little formal intersection. So the skills and technologies, the knowledge and the institutional contacts required for the circulation and preparation of city strategies certainly intersect and affect the circulation of ideas and influences. But cities also find themselves tied into certain discrete circuits of policy: for example, national legislation, EU-wide criteria for urban funding, bilateral or multilateral opportunities for international development aid and advice, or perhaps a temporary movement or network based on cities perceived to share a common problem or ambition (for example, informality, corruption in governance, commitment to environmental sustainability).

In the uptake and preparations of comprehensive city strategies will be many different policy elements, shaped by the different areas of city management under consideration, with often divergent journeys and influences. In terms of the overall policy technique, though, it is possible to build up a picture of some of the circuits and sites of influence and innovation. Some direct and easily identifiable lines of influence are from U.S. competitive city policies to the City Development Strategy of Cities Alliance, for example, via World Bank analysts and academic consultants (Harris 2002; Robinson 2008), or from participatory governance in Porto Alegre to strongly consultative city planning and strategy formation in South Africa and elsewhere (Ballard et al. 2007). There are also nonspecifiable, overlapping, and even unlikely influences at work. Private sector consultants whose international reputation gives them the profile to promote a certain approach to city strategizing (such as the Boston-based The Monitor Group did in Johannesburg) or local councils that commit to developing more local expertise in scoping city futures (as eThekwini council did in Durban) are both part of the complex circulations of ideas, expertise, and technologies that frame the field. And so are the many direct city-to-city knowledge transfers, for example, through study tours, or twinning, which took place between officials and councilors from Lusaka, Zambia, and Dayton, Ohio, which produced a version of a city strategy in Lusaka in 1999 (Lusaka City Council 1999). More important, informal exchanges; meetings at conferences; networking by businesses, citizens, city executives, and councilors; visits by academics; and books and reports received and read (or not) all add up to shaping a city strategy, affecting who will be involved in producing it, and what its outputs and achievements might be. To assemble all the lines of policy influence for even one city strategy would be incredibly difficult, although significant delineations of powerful discourses and framing approaches can be achieved and explain a lot about any particular city strategy (and see Healey [2007] for a detailed study of three cities' engagements with strategic planning over several decades as an example of the value in taking the complexities of layers of [planning] discourse seriously).

Overall, then, the case of city strategies indicates the need to interrogate the spatial metaphors at work in interpreting global policy circulations. An excellent paper by anthropologist Anna Tsing (2000) offers a useful starting point, encapsulating some of the issues involved in considering global "flows." She offers an insightful critique of the "charisma" of globalization, assumptions of global "newness," and the tendency among anthropologists and other commentators on globalization to focus on the achievement of interconnection and mobility at a global scale to the neglect of fragmentations, disconnections, and stagnations. She argues, along with many geographers, against a counterposition of the local and the global. Distinctively, she sets herself against the alignment of the global and mobile with the included, connected world and the local or immobile with excluded and marginal groups. Very helpfully, she asks the question, is globalism like modernism? With that lens she convincingly argues against transmuting the modernist tendency to cast some groups as backward, outside of the modern, into globalism's exclusions on the basis of connectedness. In a similar way to theorists of modernity (see Mitchell and Abu-Lughod 1993), she suggests that globalization is best thought of as a heterogeneous collection of diverse projects involved in making both places and mobilities, rather than as an inevitable and seamlessly unitary project associated only with interconnections.

Her spatial analysis of globalization focuses on a critique of the term *circulations*. She suggests that "we might overvalorise connection and circulation rather than attending to the shifting, contested making of channels and landscape elements" (Tsing 2000, 336). For her, interconnections are central to most accounts of globalization, and since these are created through circulations, her spatial critique of the globalization literature focuses on how the term *circulation* has been deployed. She comments that "a focus on circulations shows us the movement of people, things, ideas, or institutions, but what it does not show is how this movement depends on defining tracks and grounds or scales and units of agency . . . If we imagined creeks, perhaps the model would be different; we might notice the channel as well as the water moving" (337).

Helpfully, for our purposes, she illustrates her answer using Saskia Sassen's (2001) account of the global city, in which Sassen demonstrates how the flows of global finance are managed and enabled in specific locations in certain cities—the global flows are intimately dependent on certain territorializations of economic activity and infrastructure. In a maneuver that instinctively follows Massey (2005), Tsing (2000) observes no distinction, then, between the local and the global and directs our attention to the many different elements of place making and travel that compose contemporary (and historical) societies, and to the "missed encounters, clashes, misfires, and confusions that are as much part of global linkages as simple flow" (338).

In terms of the spaces of globalization, Tsing directs our attention away from pure flow and insists that we should pay attention to the ways in which mobilities and flows are produced and enabled and follow specific paths or tracks. She observes that "the channel-making activity of circulation, then, is always a contested and tentative formation of scales and landscapes" (338). She diverts our attention away from the connections and mobilities to a familiar topographical imagination of the spaces that generate and subtend connections, or through which they travel, or the scales that are politically and imaginatively produced through contestations over flows.

Recent geographical thinking on space asks us to press this analysis further. I feel that we can ask the term *circulations* to do rather more analytical work for us than Tsing allows. Her attempt to pull flows and interconnections—what she calls "circulations"—back to the easily mapped topographical phenomena of channels, landscapes, and scale displaces precisely what is most interesting about circulations: their often messy, unmappable complexity, and their dependence on spatialities of proximity and presence that are, again, not easily reduced to physical flows that can be traced on a map. Circulations, then, are not specifically about traversing physical distance or traceable connections but might often be more convincingly understood through the concept of topological space[1] (Amin 2002; Allen 2003, 2008). Rather than pulling us back to the physical channels and points of physical copresence that enable flows (although these, as with the details of trajectories, are certainly relevant and interesting), topological accounts of space direct us to attend to the specific spatialities at work in the drawing of people, ideas, and activities into proximity, into closer relationships, or not (Allen 2008).

In policy circulations, then, one has something altogether organization-
ally looser (untraceable, even) than the heterogeneous networks pursued by
Actor Network theorists or even the generic relational geographies that subtend contemporary accounts of space. Following John Allen (2003, 2008),
one needs to look beyond geometrical or networked metaphors to appreciate how space matters to the exercise of power at a distance. He considers
a range of spatialities—of reach, influence, drawing closer, keeping at a
distance—that operate to shape fields of power. Spatial relations of proximity and presence, then, can be seen to operate largely independently of the
channels and landscapes within which Tsing wishes to fix our unruly sense
of the interconnectedness of many elements of contemporary life.

Pierre-Yves Saunier (2002) helpfully brings together some of the ways
that international circulations of urban policy have been effected for over
a century. He sets out a definition of connections in relation to municipal policy as "a series of linkages—formal and informal, permanent or
ephemeral—which bind together entities that are geographically far apart,
either in a single country or across boundaries" (512). These connections,
which he suggests began to increase at the end of the nineteenth century,
embrace formal organizations and associations as well as individual relationships, shared publications, and targeted learning across localities. As
he points out, the field of urban policy has been strongly internationalized for more than a century, suggesting that there is much work to be
done to rewrite the many overly nationalist narratives of urban government in wealthier country contexts. He makes an important critique of the
enterprise of comparative urban studies, which occludes this fertile and
important world of connections in what remains a strongly territorialized
methodological imagination.

It might be possible to produce a snapshot of the topographical world
of flows and territorializations that urban policy traces across the globe,
mapping where ideas or practices came from, the sites where they might
be generated, where they went or returned to, although the sheer complexity of this in the contemporary world would be overwhelming: it
might look a bit like a map of the Internet! Rather than this, two observations I take from Saunier are (1) to appreciate the multidirectionality
and diversity of arenas in which international policy learning and innovation take place and to wonder to what extent and with what effectiveness
particular policy ideas can gain prominence in those circumstances; and
(2) to focus on the topological spatialities and power relationships that
enable such policy engagements and that are their outcome.

Rather than descriptions of the topographical spaces of flow and connection, then, we can be drawn to explore the spatial and power-laden

processes that support certain proximities or distancing in the world of policy formation, which enable certain ideas to be invented, put into motion, and appropriated or not. Ideas and practices move about the world in all sorts of ways—through meetings, publications, reports, scholarly writing, friendship and collegial networks, and formal institutions. The intellectually and politically important questions are, what enables ideas to take hold, connections to be forged, relationships to be formed, municipalities to pursue certain agendas, experiences to be packaged as best practice, and what are the effects of these achievements? This matters politically for two reasons. First, humanitarian concern about urban conditions in most of the world means that policy action in urban development needs to be effective. Second, concerns about the ambitions of powerful agents in this field make determining an appropriate political engagement with apparently hegemonic urban policy important. In pursuing these questions, it is my hunch that the concept of global governmentality has much to offer, and the following section considers this.

Globalizing Urban Governmentality

With a broad academic consensus that there has been a neoliberal turn in urban policy internationally comes a temptation to assess localized urban policy developments as conforming to international trends. It is too easy to see the politics of policy formulation as pitting relatively progressive forces in localities against external and more powerful actors. The power of knowledge (and often financing) is understood to be aligned with international agencies and powerful states advocating neoliberalism, and with local elites drawing on these forces to foster socially conservative policies and economic competitiveness agendas in the disciplining arena of international capital mobility (Smith 2002; Davis 2006; Harvey 2006). Alternatively, some writers invite us to take a fresh look at the relations of governance that frame international policy circulation, directing our attention to the technologies and practices of "global governmentality" and encouraging a more nuanced assessment of specific international policy movements (Larner and Walters 2004; Mosse and Bell 2005; McCann 2011). Aligned with this is a relatively open attitude to the political meaning of neoliberalism.

Neoliberal techniques of governance, Larner (2000) insists, can be deployed within a range of different conservative and progressive political projects. Neoliberalism as a policy agenda, for example, can be shaped by oppositional and democratic voices active within specific contexts. In New Zealand, which Larner has explored in some detail, feminist, aboriginal,

and social movements contributed to shaping a range of neoliberal practices that coexisted with commitments to a redistributive welfare state.
Peck (2004) has developed this insight more broadly. He sets out a strong
argument that neoliberalism is as much a product of the periphery as
the center. Chile and Mexico figure strongly in origin narratives of neoliberalism. More than this, he suggests, in the same vein as Larner, that
neoliberalism everywhere is a hybrid form, a product of particular historical and contextual forces. Thus, he proposes a strongly postcolonial
interpretation of neoliberalism, one that eschews the idea that U.S. (neo)
liberalism is the model or origin of global neoliberalism and that leaves
the political meaning of neoliberal policy adoption open to determination
in particular contexts.

Both Foucauldian and political economy approaches (here, both Larner
and Peck) lead one to assess that transnational global policies—such as
city strategies—are highly contingent, dependent on local circumstances,
and that their specific form is the result of the range of influences and connections at work in any given instance. These analyses also indicate that
traveling policy discourses remain in progress, open to contestation and
reimagination, often escaping the intentions of their architects or authors.
Whereas this may seem like good news to political activists interested in
securing progressive gains in an often unfriendly environment of state
rollback or privatization, the other side of this analysis of the power relations of international neoliberal governance is less heartening. Turning
the question around, then, one could consider how it is that organizations committed to circulating and implementing policies and practices
that might improve the effectiveness of local governments, or the lives of
urban residents, can hope to succeed when the dynamics of policy transfer
make policy take-up relatively unpredictable. Given the extent of the challenges faced by rapidly urbanizing cities in poorer country contexts, the
question as to how better service delivery and urban management practices can be embedded in poorer cities is a pressing one.

Nonetheless, in the spirit of Peck's and Larner's arguments, and in the
wake of a wider Foucault-inspired governmentality literature, we can
build on the idea that the processes of policy adoption are as likely to be
positive and enabling, creating new opportunities in and for cities, as they
might be imposed and prescriptive (Foucault 1980). Flowing from this,
the power relationships involved in policy circulation can be understood
to stretch across powers of seduction and persuasion, rather than, or as
well as, domination and delimitation (Allen 2003). This opens up space
for imagining a range of different kinds of local engagements with circuits
of urban policy and establishes the need for more subtle theorizations of

the power relations involved in circulating urban policies (see, for example, McCann 2011).

The narrative of neoliberal policy circulation associated with relatively poor countries does not intuitively lend itself to this more positive view of power relations. The imposition of structural adjustment policies in the wake of debt crises across the global South in the 1970s speaks to a sharp power gradient, in which client countries have had little choice but to impoverish their populations and significantly undermine their autonomy in developing national economic and social policies that create open economies and significantly reduce state sectors. However, detailed research on the national-level power relations between the World Bank, donor agencies, and client countries in "postconditionality" policy environments indicates the subtlety of power relations even in such circumstances. Mutual interests (for example, in external perceptions of policy success), shared discourses, and close working relationships can make distinctive agendas between donor agencies and locally based clients hard to discern (see Harrison [2001], on Tanzania and Uganda). Thus, it is possible to expect a more subtle relationship to emerge through processes of international policy development and funding (although there is also much evidence of more radically skewed IFI/donor–client relationships; for an urban South African example, see Tomlinson [2002]). In this regard, in relation to African urban policy circuits, the challenges faced by apparently powerful organizations, such as the Cities Alliance and the World Bank in convincing many countries to place urban issues higher up their priority list, have been substantial and speak to the range of power relations potentially at work in global urban governmentality, despite superficially steep power gradients.

In a more general analytical vein, we could start from the premise that power is a complex and situated achievement—not an a priori property of more or less powerful actors (Latour 1986; Allen 2003). With this in mind, urban policy initiatives are less able to be characterized as imposed from outside; and it is probable that external ideas are more likely to gain purchase when they are seen to benefit local agents or when local agents purposefully seek them out. Even when policy initiatives are relatively coerced (e.g., through threats of withholding essential funding or the promises of substantial investments in expensive infrastructure in resource-poor contexts), local actors are nonetheless quite likely to be involved in shaping the outcomes. The subtle power relations of policy adoption and adaptation in an international context clearly deserve close attention (see, notably, McCann [2011], who helpfully sets out a broad research agenda on urban policy circulations).

In sum, the potential for a policy initiative to be mobilized to benefit different local interests and agents is theoretically strong—who exactly is able to influence policy adoption in different contexts is, of course, the stuff of local politics. With broad strategic and visioning exercises such as city strategies, mandated to engage closely with local realities and diverse constituencies, there is substantial potential for creative policy adaptation and strong interrogation of externally generated policy solutions. I'll return to this point about the political possibilities of city strategies in the conclusion. I turn first to discuss contrasting examples of city strategies to consider the power relations involved in their development and adoption. In these cases, analysts report a strong sense of local autonomy in determining strategic direction, even when local governments are apparently severely constrained by national or international forces, by lack of resources and capacity, or by apparently stylized policy approaches or packages; in all cases, a form of neoliberalization of local urban policy is involved. Much more research remains to be done on the phenomenon of city strategies, especially in poorer country contexts, but these selected examples drawn from the existing literature all indicate the subtleties of the power relations shaping circulating urban policies.

An interesting study of two cities that both embraced a strategic revisioning process involving new public management principles (also known as neoliberalism) exemplifies the idea that policy adoption might often result from the powers of seduction and persuasion, as well as from a very general engagement with broad circulating policy discourses—what the authors of the study term *transnational discourse communities*—and readily available information drawn from experiences elsewhere (Salskov-Iversen et al. 2000). Both Tijuana (a border town in Mexico) and Newham (one of the poorest boroughs in London) undertook strategic city visioning exercises to redirect their fortunes; both were out of sync with national initiatives, and both independently sought solutions and strategies from the broad suite of new public management (NPM) techniques. In Tijuana where opposition politics dominated, elites benefited from greater local autonomy accorded to localities because of changes in intergovernmental relationships, although the national government had not set a specific NPM agenda for the local state restructuring they had initiated. A poor reputation and many social challenges, but also the opportunities of proximity to the United States and the strong cross-border investment in industrial development (*maquilladoras*), drew the new political elite's attention from the time of their local victory in 1989 to questions of competitive positioning, image, representation, and strategic planning, as well as to classic NPM concerns with civic partnership and efficiency in service delivery. A newly elected council undertook a substantial strategic planning exercise to reorganize the operations and image of the city. In proposing this, they drew extensively on broader international discourses of urban management and economic growth, including engagements with international consultants Arthur Andersen and local researchers as well as the World Bank and the experiences of other cities, notably Bilbao, Spain, and San Diego, California. Distinctively, they fostered a strong commitment to participatory governance, partly influenced by the World Bank, within the broad suite of NPM measures they adopted.

In Newham in the East End of London, a labor council presided over a population with significant levels of social and economic distress; it was one of the most deprived boroughs in the country with one of the worst records on service delivery in London. From the mid-1990s, new leaders turned the council toward an agenda of revitalization, service delivery efficiency (best value), and reimagining the borough as a site of opportunity within the broader regional economy. Strong strategic planning and revisioning played an important part. Long in advance of New Labour commitments to applying new public management techniques (and quite at odds with a strongly centrally directed urban policy environment) they thoroughly restructured their organization and revisioned their place in the wider London region, seeking opportunities to join development partnerships across the city.

Rather than national or international imposition of policy norms, then, local government leadership in both localities sought out ideas that they felt would work well. They embraced and adapted these to their own strategic ambitions, which included both economic success and the more effective delivery of social support. Despite the UK's position at the forefront of neoliberalization initiatives, "in Newham's case, the strategic embrace of NPM was and remains very much the responsibility of the leading councilors and chief officers; it was never imposed from above by central government" (Salskov-Iversen et al. 2000, 25). The authors conclude that their case studies of NPM adoption in Tijuana and Newham illuminate how global discourses undergo local translation, how they are contextually contingent in their reproduction, and how, in terms of global policies, the local functions as a "generating site" (26).

To study a global phenomenon such as city strategies, it is important to assess city strategy development across cities in wealthier and poorer country contexts, as Salskov-Iversen et al. do. One issue that such comparative work raises is the question of potentially different power relations between localities and external forces (national and international)—poorer localities may be considered more beholden to external parameters than wealthier contexts. Turning from Newham as representing a borough with deep social challenges to consider the whole city of London, which is usually placed at the top of global urban hierarchies, it is interesting to consider this city alongside the largest city in one of the poorest countries in the world, Dar es Salaam, Tanzania. In terms of the power relations of international urban policy adoption, one would expect a powerful city like London to have substantial autonomy in determining its strategic visions and for a city like Dar es Salaam, with few resources, to be effectively hostage to powerful international agents. However, in both cases, understanding the politics of city strategy formation requires a careful assessment of the range of different kinds of power relations at work.

In London, far from having substantial autonomy, commentators agree that the recently established Greater London Assembly and the Mayor of London have highly circumscribed ability to shape the city—although these have been enhanced recently (Thornley 2001; Travers 2004). The national government mandates London's form of strategic planning. Furthermore, much of the discretion for funding the development of the city is held by national government or at a borough level. An effective mayoralty depends on negotiation with and persuasion of competing power centers (including, of course, the powerful international and national business lobby in the city) to generate support for common strategic ambitions. Within these constraints, the city strategy developed under the first mayor of London, Ken Livingstone, managed to mobilize key constituencies toward a long-term spatial plan envisaging extensive eastward expansion and leveraging central government support for the major infrastructural development this would entail (coinciding usefully with the successful Olympic bid for 2012). Thus, a relatively powerless mayor, through subtle mechanisms of influence, positioning, and alliance-building in the process of strategic visioning, created a strong and politically viable growth strategy for the city (Gordon 2003).

In Dar es Salaam, similar constraints on autonomy and financing hobbled local-level initiatives. A city strategy process, focusing on long-term planning for the UN Environment Program's Sustainable Cities Program, was implemented in 1999 through the national government's interface with international donors, thus potentially bypassing local leaders and community members (this example is drawn from Myers [2005]). It nonetheless generated enthusiasm across a number of interest groups and led to substantial changes in local government practice, especially regarding solid waste disposal. Its success was also partly due to a charismatic local leader who used a window of political opportunity occasioned by integrated metropolitan government to advance the process and to achieve local development priorities. However, national-level opposition to the relatively powerful political platform afforded by integrated metropolitan government saw the city's government fragmented once more, and the broad strategic thrust of the initiative was lost. Nonetheless, Myers (2005) reports on the subtle ability of actors at the local state level to mobilize, divert, and even sideline the externally imposed and donor-funded agendas that were articulated through the city strategy process—in one example, quite literally so in terms of the relatively inaccessible physical location of those working on a key part of the strategic program. Both the London and Dar es Salaam examples remind us to look closely at the specific, multidimensional power relations of policy circulations, rather than assume a single narrative on the basis of either relative political or economic power.

The case of Johannesburg, South Africa, speaks even more powerfully to the importance of local experiences in shaping a policy initiative (see Lipietz 2008). There, the CDS was conceived and initiated locally and followed a well-trodden postapartheid path of bringing "stakeholders" to the negotiating table. It articulated a strong antiapartheid discourse in which the city was to be addressed as a whole, cutting across inherited racial and class divides. Senior officials and councilors hint that they invented CDS, which was later taken up by the World Bank and the Cities Alliance—although both former World Bank president Wolfensehn (2001) and the Cities Alliance claim, to the contrary, that Johannesburg was the first place to implement their new CDS strategy. Either way, this policy technique found a strong resonance in a postapartheid context and has since been mainstreamed in national policy guidance and promoted by the South African Cities Network, spanning the nine major metropolitan areas (Harrison 2006). A number of CDSs have been conducted in South Africa, interestingly with a wide range of different policy conclusions and political ambitions despite the common national policy and economic context the cities there face. What might appear to be an instance of the local application of global policy discourse in the Johannesburg case was a strongly locally determined policy process, shaped by quite specific political dynamics.

The available evidence, then, suggests that global urban policy, developed through the loose and relatively unpredictable networks of political influence, policy circulation, and financial enticement, operates in a field of power relations that is helpfully characterized by the governmentality literature. In this perspective, there is scope for mutuality in the definition of policy discourses and agendas, certainly scope for transformation of the political meaning of techniques of governing, and even scope for the local invention and delineation of new policy directions.

Conclusion: The Local Potential of Global Policies

Theorizing the mobility of urban policy means paying attention to the range of sites and the diversity of tracks that compose a globalized field of urban development knowledge. This includes the dense associations represented by agencies and donor groupings, or by the networks of consultants and theorists as well as the city-specific, territorialized agents involved in policy circulation. It also means paying attention to the topological spaces of policy adoption, for as we have seen, circulation happens through active appropriation; it is characterized by dynamic engagements of territorially committed actors with ideas and practices from within and beyond their particular locality. Attending to the subtleties of the power relations involved in globalized circuits of urban knowledge should therefore cause us to question any a priori expectation that local political actors might necessarily align themselves for or against a neoliberal policy agenda. In relation to the wider circuits of knowledge, then, the previous discussion suggests that the potential for local agency to appropriate and creatively use (even mandate) opportunities for citywide strategic planning has arguably been understated, especially in discussing the circulation of neoliberal policy.

However, it is important to place this relatively enabling analysis alongside careful assessments of the lineaments of the local politics that shape city strategies. While Raco et al. (2006) confirm the scope for maneuver in relation to strategic policy outcomes in their study of Community Strategies (CS) in the UK, others point to some difficult and entrenched local-level institutional dynamics that might generally be considered to limit the potential of city strategies. They assess the relative significance of representative (electoral) and participatory democracy as opposed to bureaucratic direction in the preparation of CSs (see Ballard et al. [2007] for a comparable assessment of CDS in eThekwini Municipality, in Durban, South Africa). Considering the complex interplay among these diverse elements of local government in two different areas of the UK, they conclude that "there is no one trajectory of CS development and that the effects of strategy implementation cannot be surmised a priori" (Raco et al. 2006, 493). Other studies point to the ways in which institutional, political, and broader contextual factors matter and might limit the possibilities for relatively productive outcomes of strategic planning—as studies by McCann (2001) in the United States and Geddes (2006) in the

UK conclude. Inadequate participation, strong elite control of planning processes, and substantial limits to local autonomy in centrally driven political regimes can all direct or mobilize city strategies toward nonprogressive ends. As Geddes (2006) concludes of the UK case,

> the emerging evidence is that, while local partnerships might—in a different context—have the potential to enhance local democracy, help rebalance the central-local relationship and improve governance effectiveness and outcomes, the concern must be that they undermine democracy and accountability and lack the power and capacity to be effective, while limiting local policy options to those consistent with the neoliberal agenda which dominates New Labour public policy. (93)

However, the wider comparative literature that we have discussed does not confirm his more general assessment that "to change this would require not (just) better local governance arrangements, but a rejection of neoliberalism as the basis of public policy" (93). Indeed, even the achievement of managerial and central control of a policy process is exposed to challenge and rests on the subtle, uncertain power relations that the governmentality literature excavates in its analysis of neoliberalism. As Lever (2005) suggests, central control depends on relatively unpredictable and ultimately subvertible process of adopting national and international policy discourses to achieve local ends. He reports "communities internalizing the policy discourses of government and developing technologies of government that allow them to attract mainstream funds" (912).

Plenty of cross-national evidence indicates that building creative, locally generative strategic visions and comprehensive plans for city futures is possible, even within a broadly neoliberal international policy environment. However, this requires substantial further comparative reflection, as it must often take place in city contexts riven with unequal power relations. In addition, and important for any initiative aiming to extend city strategizing, with the limited capabilities for strategic city-wide visioning, especially among community groups, and often difficult economic and developmental trajectories to contend with, this policy technology faces some substantial obstacles. Nonetheless, this chapter has suggested that thinking through the spatialities and power relations of global urban governmentality indicates that the potential does exist for city strategies to act as sites of opportunity for reformulating and redirecting local futures.

Note

1. Topology: In mathematics, this refers to spatial relationships among features, including adjacency and connectivity. It implies relative position, as opposed to absolute position specified by coordinates, angles, and distances.

References

Allen, J. 2003. *Lost Geographies of Power*. Oxford: Blackwell.

———. 2008. "Powerful Geographies: Spatial Shifts in the Architecture of Globalization." In *Handbook of Power*, edited by S. Clegg and M. Haugaard, 157–74. Oxford: Oxford University Press.

Amin, N. 2002. "Spatialities of Globalization." *Environment and Planning A* 34:385–99.

Asian Development Bank. 2004. *City Development Strategies to Reduce Poverty*. Manila: Asian Development Bank.

Ballard, R., D. Bonnin, J. Robinson, and T. Xaba. 2007. "Development and New Forms of Democracy in Durban." *Urban Forum* 18:265–87.

Cities Alliance. 2006a. *Guide to City Development Strategies: Improving Urban Performance*. Washington, D.C.: Cities Alliance.

———. 2006b. *2006 Annual Report*. Washington, D.C.: Cities Alliance.

Cochrane, A. 2006. *Understanding Urban Policy: A Critical Approach*. Oxford: Blackwell.

Connell, R. 2007. *Southern Theory: The Global Dynamics of Knowledge in Social Science*. Cambridge: Polity Press.

Cox, K. R., and A. Mair. 1988. "Locality and Community in the Politics of Local Economic Development." *Annals of the Association of American Geographers* 78:307–25.

Davis, M. 2006. *Planet of Slums*. London: Verso.

Foucault, M. 1980. *Power/Knowledge: Selected Interviews and Other Writings, 1972–1977*. Edited by C. Gordon. Brighton, UK: Harvester Press.

Geddes, M. 2006. "Partnership and the Limits to Local Governance in England: Institutionalist Analysis and Neoliberalism." *International Journal of Urban and Regional Research* 30, no. 1: 76–97.

GHK Consultants. 2002. "City Development Strategies: An Instrument for Poverty Reduction?" Final report presented to Department for International Development, UK, August.

Gordon, I. 2003. "Capital Needs, Capital Growth, and Global City Rhetoric in Mayor Livingstone's London Plan." Paper presented at the annual meeting of the Association of American Geographers, New Orleans, La., March.

Harding, A. 1994. "Urban Regimes and Growth Machines: Towards a Cross-National Research Agenda." *Urban Affairs Quarterly* 29, no. 3: 356–82.

Harris, N. 1995. "Bombay in a Global Economy: Structural Adjustment and the Role of Cities." *Cities* 12:175–84.

———. 2002. "Cities as Economic Development Tools." *Urban Brief*. Washington, D.C.: Woodrow Wilson International Center for Scholars.

Harrison, G. 2001. "Post-Conditionality Politics and Administrative Reform: Reflections on the Cases of Uganda and Tanzania." *Development and Change* 32:657–79.

Harrison, P. 2006. "Integrated Development Plans and Third Way Politics." In *Democracy and Delivery: Urban Policy in South Africa*, edited by U. Pillay, R. Tomlinson, and J. du Toit, 186–207. Cape Town, South Africa: HSRC Press.

Harvey, D. 2006. *Spaces of Global Capitalism*. London: Verso.

Healey, P. 2007. *Urban Complexity and Spatial Strategies: Towards a Relational Planning for Our Times*. London: Routledge.

Jessop, B., and N. Sum. 2000. "An Entrepreneurial City in Action: Hong Kong's Emerging Strategies in and for (Inter)Urban Competition." *Urban Studies* 37:2287–313.

Larner, W. 2000. "Neo-Liberalism: Policy, Ideology, Governmentality." *Studies in Political Economy* 63:5–26.

Larner, W., and W. Walters, eds. 2004. *Global Governmentality: Governing International Spaces*. London: Routledge.

Latour, B. 1986. "The Powers of Association." In *Power, Action, and Belief*, edited by J. Law, 264–80. London: Routledge and Kegan Paul.

Lever, J. 2005. "Governmentalisation and Local Strategic Partnerships: Whose Priorities?" *Environment and Planning C: Government and Policy* 23:907–22.

Lipietz, B. 2008. "Building a Vision for the Post-Apartheid City: What Role for Participation in Johannesburg's City Development Strategy?" *International Journal of Urban and Regional Research* 32:135–63.

Logan, J., and H. Molotch. 1987. *Urban Fortunes: The Political Economy of Place*. Berkeley: University of California Press.

Lusaka City Council. 1999. *Five Year Strategic Plan*. Lusaka: Lusaka City Council.

Machimura, T. 1998. "Symbolic Use of Globalisation in Urban Politics in Tokyo." *International Journal of Urban and Regional Research* 30:183–94.

Massey, D. 2005. *For Space*. London: Sage.

McCann, E. J. 2001. "Collaborative Visioning or Urban Planning as Therapy: The Politics of Public–Private Urban Policy Making." *Professional Geographer* 53, no. 2: 207–18.

———. 2011. "Urban Policy Mobilities and Global Circuits of Knowledge: Towards a Research Agenda. *Annals of the Association of American Geographers* 101, no. 1:107–30.

Mitchell, T., and L. Abu-Lughod. 1993. "Questions of Modernity." *Items* 47:79–83.

Mosse, D., and D. Bell. 2005. *The Aid Effect: Giving and Governing in International Development*. London: Pluto.

Myers, G. 2005. *Disposable Cities: Garbage, Governance, and Sustainable Development in Urban Africa*. Aldershot: Ashgate.

Parnell, S., and J. Robinson. 2006. "Development and Urban Policy: Johannesburg's City Development Strategy." *Urban Studies* 43, no. 2: 337–55.

Peck, J. 2004. "Geography and Public Policy: Constructions of Neoliberalism." *Progress in Human Geography* 28, no. 3: 392–405.

Pieterse, E. 2008. *City Futures: Confronting the Crisis of Urban Development*. London: Zed Books.

Raco, M. 2007. *Building Sustainable Communities: Spatial Development, Citizenship, and Labour Market Engineering in Post-War Britain*. Bristol, UK: Policy.

Raco, M., G. Parker, and J. Doak. 2006. "Reshaping Spaces of Local Governance? Community Strategies and the Modernisation of Local Government in England." *Environment and Planning C: Government and Policy* 24:475–96.

Robinson, J. 2006. *Ordinary Cities: Between Modernity and Development*. London: Routledge.

———. 2008. "Developing Ordinary Cities: City Visioning Processes in Durban and Johannesburg." *Environment and Planning A* 40:74–87.

Robinson, J., and C. Boldogh. 1994. "Local Economic Development Initiatives in the Durban Functional Region." In *Local Economic Development in South Africa*, edited by R. Tomlinson, 191–214. Johannesburg, South Africa: Witwatersrand University Press.

Salskov-Iversen, D., H. K. Hansen, and S. Bislev. 2000. "Governmentality, Globalization, and Local Practice: Transformations of a Hegemonic Discourse." *Alternatives: Global, Local, Political* 25, no. 2: 183–223.

Sassen, S. 2001. *The Global City: New York, London, Tokyo*. Princeton, N.J.: Princeton University Press.

Saunier, P.-Y. 2002. "Taking Up the Bet on Connections: A Municipal Contribution." *Contemporary European History* 11, no. 4: 507–27.

Smith, N. 2002. "New Globalism, New Urbanism: Gentrification as Global Urban Strategy." *Antipode* 34, no. 3: 427–50.

Stren, R. 2001. "Local Governance and Social Diversity in the Developing World: New Challenges for Globalising City-Regions." In *Global City-Regions: Trends, Theory, Policy*, edited by A. J. Scott, 193–213. Oxford: Oxford University Press.

Thornley, A. 2001. "Agenda-Setting in the GLA with Particular Reference to City Marketing and the Role of the LDA." In *Governing London: Competitiveness and Regeneration for a Global City*, edited by S. Syret and R. Baldock, 68–81. London: Middlesex University Press.

Tomlinson, R. 2002. "International Best Practice, Enabling Frameworks, and the Policy Process: A South African Case Study." *International Journal of Urban and Regional Research* 26, no. 2: 377–88.

Travers, T. 2004. *The Politics of London: Governing an Ungovernable City.* Basingstoke, UK: Palgrave Macmillan.

Tsing, A. 2000. "The Global Situation." *Cultural Anthropology* 15, no. 3: 327–60.

United Nations. 2006. *State of the World's Cities 2006/7: The Millennium Development Goals and Urban Sustainability.* New York: UN-Habitat.

Vicari, S., and H. Molotch. 1990. "Building Milan: Alternative Machines of Growth." *International Journal of Urban and Regional Research* 14:602–24.

Ward, K. 2006. "'Policies in Motion,' Urban Management, and State Restructuring: The Trans-Local Expansion of Business Improvement Districts." *International Journal of Urban and Regional Research* 30, no. 1: 54–75.

Wolfensohn, J. 2001. "The World Bank and Global City-Regions." In *Global City-Regions: Trends, Theory, Policy*, edited by A. J. Scott, 44–49. Oxford: Oxford University Press.

World Bank. 2000. "Cities in Transition: World Bank Urban and Local Government Strategy." Washington, D.C.: World Bank.

CHAPTER THREE

Creative Moments

Working Culture, through Municipal
Socialism and Neoliberal Urbanism

Jamie Peck

Introduction: Mobilizing Urban Culture

Policies designed to stimulate the "creative growth" of city economies—
usually by way of market-friendly interventions in the cultural sphere,
to attract or retain elite workers—might be characterized as the most
conspicuously successful innovation in the recent history of urban policymaking. They are "successful" in the sense that the reach of these policies
seems to have become near ubiquitous, even if they are apparently spreading by way of increasingly pale imitation (Peck 2005; Ross 2006; Scott
2006). Catering to the creative class has become, almost simultaneously,
a favored strategy, an urgent imperative, and a hackneyed cliché of contemporary urban policymaking. International in scope, the market for
creativity policies now extends from the top to the bottom of the urban
hierarchy, animated by a pervasive sense of competitive urgency: in an era
of knowledge-driven growth, an influential thesis has it that productive
potential is carried by a creative *class* of individuals, who will be attracted
(only) to cities with "buzz," cities with a welcoming and sustaining "people climate" (Florida 2002, 2005, 2008). The idealized subjects of this
new urban economy, a hypermobile elite of high-tech hipsters, allegedly
crave opportunities to maximize their innate talents in the context of 24/7
experiential intensity. The now-storied creative class inhabits a socially
and economically liberalized world, in which the barriers to sociospatial
mobility are progressively removed. And its members are the objects of
accelerating "talent wars," among both corporations and cities.

Extant urban development models—based, inter alia, on the aggressive pursuit of investment opportunities, city marketing, supply-side inducements, property-led growth, gentrification, commodified culture, and retail revitalization—have been retrofitted around this new vision of creative growth. None of this was a huge stretch. The established strategy of manipulating business climates to attract mobile investment has effectively been overlaid with the new "creative" approach, of manipulating people climates to attract mobile talent workers. In this new iteration of the urban development game, a refusal to play practically invites competitive failure, at least according to the prevailing script. Buzzing cities can anticipate cumulative growth, but staid, hierarchical, and suburbo-centric cities, where the buzz is *off*, can only slide into the rustbelt of the knowledge economy. However, playing is both easy and relatively cheap, while the distributors of the new dispensation insist that *every* city has a shot at winning, as long as policy elites are capable of mastering Richard Florida's alliterative trinity of technology, talent, and tolerance. Creativity has ostensibly become the new (universal) formula for urban growth, just as its accompanying policy routines—of culturally inflected economic development, rebadged promotional strategies, and new age gentrification—have become decidedly formulaic (chapter 4).

What should be made of this? Is it a mere policymaking "fad"—as transitory as it is trivial? Appropriately, perhaps, "entrepreneurialized" city managers seem to have become just as susceptible to fads as their competitively insecure colleagues in the corporate world. But managerial fads, as Abrahamson (1996) and others have noted, may be many things, but trivial they are not. The appetite for fads reflects the anxieties of managing, with limited capacities, in an uncertain environment. The market for urban policy fads happens also to have grown in lockstep with the intensification of interurban competition, and with the shared culture of existential insecurity that this has fostered among city "managers." The demand for urban bromide, in this sense, reveals more about the psychological needs of the consumers than the (supposed) efficacy of the "product." This may explain why Scott (2006, 4) chooses to characterize the apparent state of dependency on creativity fixes and instrumentalized cultural policies in the apt language of a "syndrome." As the syndrome has been transmitted, through viral urban networks, policy quacks of all kinds—from gurus to consultants—have been doing brisk business. And new urban policy products have been flying off the shelves.

Creative cities strategies—stylized as readily consumable *models*—have been tailored to these market conditions. Larded by an immodest sales pitch, they are framed as enabling policy technologies and packaged in essentialized form, to facilitate their portability from place to place and their adaptability to a range of local conditions (see chapters 2, 4, and 5 for other examples of policies in motion). The reach of these market-like systems has been massively extended by the paraphernalia of new communications technologies, from international policy conferences to Web-based resource banks, together with a host of marketing, promotional, and consulting techniques imported from the private sector. An increasingly reflexive network of "fast policy" circuits has been established, in recent years, between cosmopolitan sites of policy learning, experimentation, and emulation (see Peck and Theodore 2010). This represents more than a heightened propensity for cities to "borrow" policy ideas and techniques from one another, although this has certainly been a notable empirical trend. It signals conditions of *endemic policy mobility*, the characteristics of which include growing deference to "best practice" models, near continuous learning from like-minded (and ideologically aligned) others, and the blurring of jurisdictional boundaries. In effect, the urban policy*making* process has become increasingly relativized; it has become an *inter*urban process.

To understand what is distinctive—historically, ideologically, and practically—about today's fast policy market for creativity makeovers, this chapter invokes a critical counterpoint in the form of the short-lived wave of "cultural industries" initiatives developed by municipal socialist councils in Britain in the early 1980s. In a sense, these were the modern precursors to contemporary creativity policies, and the audacious cultural strategy of the Greater London Council (GLC) was their "seminal" moment (Hesmondhalgh 2007, 139). These pioneering cultural policies were forged under quite different circumstances, with sharply contrasting objectives. Under the vanguard influence of the GLC, cultural industries policies sought to problematize the conditions of creative production and distribution, while pioneering practice-based alternatives to neoliberalism, then in its monetarist guise. Their promise was never realized, not least because the GLC was abolished by Margaret Thatcher's Conservative government, but they indicated a radical early current in urban-cultural policy.

Here, a sketch of cultural industries policymaking in London, focused on the moment of policy *production,* is followed by an examination of the *reproduction* of contemporary creativity policies, focusing on the site of the second international creativity "summit," Detroit. This more recent moment of "cultural economy" policymaking is appropriately concerned with systemic mimicry and shallow simulation, dominant characteristics of creative cities policymaking, for all its protestations about grassroots authenticity. What follows is, then, not so much a conventional comparison of cultural industries–creative cities policies, but an attempt to draw a stylized contrast between a historical site of origin-cum-innovation, in early 1980s London, and one (of many) spaces of circulation-cum-emulation, in contemporary Detroit. Beyond the intrinsic differences in the policies mobilized in these two settings, they will each be shown to indicate distinctive policymaking milieux, together with quite different "ecologies" of policy formation and mobility. While the London and Detroit cases can both be seen as examples of local states *putting culture to work*, for the municipal socialist councils, this was part of a wider strategy of (disruptive) politicization, whereas today it is more commonly associated with a pervasive ethos of (instrumental) economization, not to say depoliticization. Moreover, while the municipal socialist strategies of the early 1980s were organically embedded in distinctive urban political cultures (see Bianchini 1989), expressing antagonistic attitudes toward the orthodoxies of both the (new) right and the (bureaucratic) left, contemporary creativity strategies are pragmatically oriented to the globalizing soft center of market-oriented politics, modestly accessorizing orthodox urban development agendas (Peck 2009). If, in retrospect, municipal socialism stands as an (ultimately frustrated) attempt to build new forms of local–state capacity, in the service of progressive political objectives, current rounds of creative cities policies threaten the continued degradation and distortion of governing capacities, in the (often futile) pursuit of the sparse compensations of neoliberal urbanism.

Culture as Work in Progress:
Municipal Socialism and the Creative Industries

The municipal socialism of local authorities like Sheffield, the West Midlands, and the GLC was a very British creation. A "new urban left" had mobilized during the 1970s as a network of (mostly big-city-based) activists and intellectuals critical not only of the ascendant project of Thatcherism but also of corporatist, bureaucratic, and reformist elements within the Labour Party (Gyford 1985). The capture and transformation of local government, as an alternative site for political mobilization and experimentation, represented a pragmatic alternative to both the parliamentary and the revolutionary roads to socialism. This targeting of local government reflected both the weakness and the potential of the British left, in this early phase of especially confrontational Thatcherism (cf. Barnett 1985). Moribund at best, the local state had previously been practically written off "as an agent of central government and handmaiden of capital" (Green 1987, 207). With roots in community development activism, feminist politics, the trade union movement, and the left of the Labour Party, the municipal socialist project sought to remake the local state as an arena of active resistance to Thatcherism, while at the same time developing working socialist models, ripe for "scaling up." In this context, what was styled as the "local economy" became a key battleground. Although it was clearly understood that the resources of local government were unequal to the daunting tasks of responding to deindustrialization and mass unemployment in a comprehensive fashion, the space was nevertheless opened up for policy interventions that might serve as "demonstration projects," as "parables" of alternative forms of socialist policy development, and as an incipient "propaganda of practice" (Ward 1983, 28; Alcock et al. 1984).

The proliferation of radical employment and economic development initiatives at the local scale was conceived, first, as a frontal challenge to the Thatcherite conceit that "there is no alternative," but second, and no less important, as a counter to the discredited forms of Keynesian corporatism, pursued by Labour governments at the national level during the 1970s. If monetarism represented a repudiation of Keynesianism from the right, mobilizing market forces as the impetus for restructuring, municipal socialism spurned it from the left, mobilizing local social and political capacities in the interests of *restructuring for labor* (Rustin 1986). The challenge was to develop new rationales and repertoires for local economic intervention, in the face of the dogmatic monetarism of the Conservative government and the late-Keynesian malaise afflicting the opposition benches. In this context, local government provided an opportunity (albeit a highly constrained one) for ideological renewal and grassroots policy experimentation (see Ward 1981; Blunkett and Green 1983). Combining modest resources with the energy and commitment of oppositional politics, local economic strategies could not realistically aspire to the material transformation of the local economy per se, but they did seek to enrich and enervate socialist *practice*, knowingly opening up opportunities for political demonstration and mobilization (see Goodwin and Duncan 1986). The process was turbulent, unruly, and disruptive. In the forefront of this effort, the GLC had "mobilised some of the most innovative political talent [available]," as Hall (1984, 39) remarked at the time, "unleash[ing] a stream of new political thinking and ideas, even though it has not always known what to do with the talent and the ideas!"

Through the creative use of its circumscribed powers of economic intervention, the GLC confronted the challenge of "restructuring and modernizing London's industrial core," through the propagation of "real working alternatives" and new practices of economic participation

(Ward, in GLC 1985, vii, ix). Through the (apparently innocuous) adoption of "detailed sectoral planning," the GLC intervened directly, but strategically, in productive and market processes, in service of the strategic goal of "reclaiming production" (GLC 1985, 18). A distant political prospect in any case, a Keynesian reflation alone would clearly not have addressed the structural problems of the London economy, which were diagnosed, in a sobering (but never debilitating) manner, in terms of deindustrialization, urban economic collapse, and the generalized deterioration of working conditions. Transformative interventions in production, in working arrangements, and in the organization of markets, including a redefinition of "work" and the development of socially useful forms of production, would be necessary in this new economics not only of, but *for*, labor. Eschewing a singular master plan, the GLC initiated a series of strategic forays into different sectors of the capital's economy, from vehicle manufacturing to software, each envisioned as a (distinctive) site of struggle and transformation, rather than merely a space of intervention (Eisenschitz and North 1986). Military metaphors were invoked explicitly, in what was seen as a multifront war, waged across a hostile terrain: building alliances with labor unions, community organizations, and social movements, a politicized cadre of municipal officers sought to engineer alternative models in the style of political agitants, rather than experts or technocrats, as "economic craft workers not economists" (GLC 1985, 2; Wainwright 1984; Mackintosh and Wainwright 1987).

Tailored strategies were developed for a wide range of "sectors," spanning both traditional sites of intervention (like clothing, instrument engineering, and information technology) and radically new fields, such as the domestic and caring economies, defense, and the cultural industries. These entailed anything but cookie-cutter replication but, instead, were grounded in a strategic (and very much *political*–economic) analysis of each individual sector. In the case of the cultural industries strategy, this went far beyond the trivial observation that cultural employment represented a significant slice of the London labor market, to embrace the character cultural commodities and production chains, the consequences of monopolization of (both public and private) distribution channels, the structure of risk, the degree of integration in cultural job markets, and the impact of new technologies. More than this, the strategy explicitly confronted the "deep-rooted antagonism towards any attempt to analyse culture as part of an economy" (GLC 1985, 172), invoking Theodor Adorno and Raymond Williams in a principled rejection of idealist and elitist notions of culture, artificially separated from the everyday, the popular, and the material (Garnham [1983] 1990; Mulgan and Warpole 1986). There was a desire to enervate and democratize culture, broadening the access to, and recognition of, a range of vernacular, popular, and working-class cultures, all emphasizing "distribution and the reaching of audiences," as opposed to "an artist-centred strategy that subsidized 'creators'" (Hesmondhalgh 2007, 139). A radical *cultural economy* policy would therefore necessarily encompass the repoliticization of both culture and economy. The once sleepy domain of local authority arts policy was reclaimed as a space for transformative investments in, and celebrations of, an entire array of working-class and minority cultures. The cultural industries policy was not only concerned with interventions in the sphere of production and production politics (bread-and-butter issues across many of the other sectoral initiatives), but more as a means to pluralize the spheres of cultural distribution and consumption, to broaden access to cultural markets and cultural work, and to recognize the creativity of marginalized social groups (Garnham [1983] 1990; Eisenschitz and North 1986). Not infrequently, these public interventions triggered conflict and controversy, though in part they were intended as disruptive policies, opening up ways in which "cultural life [might be] reconstituted as a site of politics" (Hall 1984, 39). The GLC's bottom-up approach to cultural policy, which was framed around "communities of interest," like women's and gay rights movements, the Afro-Caribbean and Irish communities, youth, and the disabled, was often associated with a turbulent politics of spectacle, for which the GLC was to be remembered long after its abolition at the hands of an implacably hostile Thatcher government (Mulgan and Warpole 1986; Hosken 2008).

Clearly, a political calculus was behind this approach as well as an economic rationale, because one of the hallmarks of the GLC's economic strategy was the commitment to work "in and against the market." In the context of cultural industries policy, this meant working "against the market's narrowing commercialising tendencies, but in the market [as] the main site where cultural needs are met or ignored, the site where jobs are created and destroyed" (GLC 1985, 185). This was translated into a program of investment in the cultural "base," emphasizing targeted assistance to independent producers and small enterprises, and a distinctive orientation to minority and alternative participation, with a view to securing stronger positions within the sphere of distribution. Echoing the Emilian "flex-spec" model, which enjoyed considerable currency at the time (Murray 1987; Nolan and O'Donnell 1987; Geddes 1988), there was also some provision of "common services," such as marketing, technical assistance, and managerial advice.

The cultural economy craftworkers at the GLC concluded that the system of cultural distribution represented "the key locus of power and profit" (Garnham [1983] 1990, 62), suggesting a strategy based less on interventions at the point of cultural production, but targeted instead on (marginally) redistributing risks, flows of reinvestment, and the management of cultural repertoires in the distributive sphere. Left to its own devices, the market was deemed "inadequate" both for satisfying cultural needs and for generating sustainable employment opportunities, but neither in principle nor in practice could a comprehensively antimarket position be sustained, given the council's limited spending power and the market's "potential for responsiveness, [its] openness to popular cultures, and its ability to reflect changing cultural needs" (GLC 1985, 184–85). Moreover, the GLC and its advisors were hardly less critical of traditional *public* interventions in the cultural realm, taking on the BBC, for example, for its elitism and its patronizing attitude to minority cultures. Classically, public funding had been directed toward maintaining "higher" forms of noncommercial culture, while the cultural needs of the masses were left to the market (GLC 1984). More fundamentally, arts and cultural policy in Britain remained fundamentally framed in idealist terms, separate from (or above) the vulgar terrains of popularization and commercialization. There was a need also to challenge the idealized, romantic notion of the artist, qua individual genius, supposedly insulated from worldly concerns and everyday working-class life (see Williams 1958, 39; Garnham [1983] 1990), in favor of a materialist understanding of cultural labor. This said, there were tensions between the GLC's *own* productivist model—based on cooperative forms of employment and clear lines of accountability—and the street-level realities of the cultural industries: in a sector heavily populated by "hustlers, entrepreneurs, sole traders, dole-queue economy wheelers and dealers," Greenhalgh et al. (1992, 128) recalled, the vision of a democratized local economy sometimes fitting "as uneasily as a Marks and Spencer polyester suit."

Ultimately, the GLC's experiment in cultural intervention was to be frustrated. As the political struggle with the Thatcher government intensified, the authority was progressively stripped of its powers in transportation, planning, and housing, while steps were also taken to rein in its capacity for economic intervention. In this context, cultural policies assumed "considerably increased . . . status within the administration" (Bianchini 1987, 43; 1989), at least until the moment of the Thatcher government's audacious abolition of the GLC, along with the metropolitan tier of government across the country, in 1985. One of the legacies of the GLC was the glimpse that it provided of a form of politics "root[ed] in the everyday

experience of popular urban life and culture" (Hall 1984, 39). Its reper-
toire of local economic strategies, which had "creatively stretch[ed] local
authority powers to their limits" but never had to face the daunting challenge of longer-term sustainability (Cochrane 1986b, 193), would later
constitute an important plank in Labour's (unsuccessful) attempt to regain
national office in 1987 (Campbell et al. 1987; Davenport et al. 1987;
Cochrane 1988).

An important purveyor of this local "jobs plan" approach, the progressive local authority think tank, the Centre for Local Economic Strategies
(CLES), had been established in 1986 with the aid of "tombstone" funding from the GLC and some of the other metropolitan counties.[1] The
impetus for the establishment of CLES was the *absence* of a

> national economic policy institute to develop new ideas and approaches,
> firmly based around the objectives of full employment, equality of
> opportunity and meeting community needs. [The] various local
> attempts to develop economic strategies and initiatives needed to be
> co-ordinated and consolidated . . . A detailed and authoritative body of
> work could then be built up with a perspective which extended beyond
> the current dominant economic theories. (CLES 1986, 1)

The phonebook-sized *London Industrial Strategy,* published in 1985, on
the eve of the GLC's abolition, was likewise intended both as a prov-
ocation to the Conservatives in Westminster and as an if-only account
of suggestive models, ready for emulation, which could be "sold to the
Labour Party [in order that it might] be implemented nationally later"
(Cochrane 1986b, 191). The continuing electoral woes of the Labour
Party at the national level (together with its rightward drift) meant that
the project was never scaled up in this way, leaving unresolved a series
of searching questions concerning its political–economic sustainability
(Nolan and O'Donnell 1987). These provocative forays into the politics of restructuring may have disrupted monetarist claims to a de facto
monopoly over policymaking rationality (see Murray 1983), while pioneering "creative forms of socialist practice" (Jessop et al. 1988, 122),
but most were to remain demonstration projects and some amounted to
little more than tokenistic gestures. Broadly sympathetic critics would
later differ on the question of the longer-term prospects of the GLC's
somewhat inchoate rendering of progressive post-Fordism, supply-side
socialism, and restructuring for labor (see Cochrane 1986a, 1986b, 1988;
Eisenschitz and North 1986; Rustin 1986; Nolan and O'Donnell 1987;
Geddes 1988; Graham 1992). For Geddes (1988, 95), this incomplete
vision of "left post-Fordism" threatened to degrade into a resuscitated

form of Keynesianism, into local corporatism, or "into a populism in which 'anything goes,' as in the GLC's strategy for cultural industries." In Eisenschitz and North's (1986) deconstruction of the London Industrial Strategy, the embryonic cultural industries effort was cautiously classified, alongside other *redistributive* initiatives, as a measure primarily aimed at positive discrimination in the job market.

It is undeniably the case that some of the claims in the London Industrial Strategy were hyperbolic, some merely aspirational (see Cochrane 1986b). But it reflected an approach of inventively targeting *politically* strategic interstices within a restructuring economy. This was not some naïve plan to construct a "metropolitan-level elastoplast," in the face of global restructuring and monetarist (mis)management, but instead was conceived as a series of political incursions: "Although the material power was really not so great at all," Massey (1997, 159) later reflected, "the symbolic impact could be considerable" (and see chapter 1, for a contemporary case of the symbolic politics of policymaking with links to the GLC's legacy). The provisional notion of a cultural industry that was folded into this broader political vision bears little relationship with contemporary visions of the creative economy as a site of self-sustaining, superior, and self-organizing forms of growth. Rather, it was seen as an underappreciated location of potential employment opportunities for marginalized social groups, one enmeshed in the "practice of everyday life." Significantly, it was also a space of representation and recognition that could be animated by multiple "communities of interest" (GLC 1985, 169). As a form of economic intervention, the cultural industries strategy sought to work against tendencies for the monopolization and industrialization of culture. The analysis was sober and materialist, refusing to idealize cultural work and workers (Garnham [1983] 1990). It sought explicitly to problematize, and grapple with, cultural markets and cultural work, investing selectively in what was represented as the cultural *base*, to improve the positions of groups underrepresented in the formal cultural economies of the public and private sectors.

In policy development terms, the energy and inventiveness of the first half of the 1980s were never regained, drifting toward a "new realist," proto–New Labour consensus after the Conservative general election victory of 1987. For a while, organizations like CLES sought to keep the flame alive, enriching the local economic policy capacities of municipal authorities "from the outside," through training and information dissemination activities, through forms of progressive consulting, and through the collaborative sharing of innovative practice, based on "working models." There was some networking across the remaining cultural industries initiatives, albeit tempered with an awareness that the boldest experiments had often been the product of the distinctive, *local* political cultures of cities such as Sheffield, London, and Manchester (see Alcock and Lee 1981; Ward 1983; Quilley 2000; Moss 2002). However, the more "transferable" lessons—almost by definition, less disruptive and radical—began to merge and meld with mainstream urban regeneration approaches (see Bailey 1989; Greenhalgh et al. 1992). CLES's position, too, would drift closer to the pragmatic center.

During the late 1980s, cultural industry themes were gradually reworked into broader strategies for downtown revitalization, retail diversification, and urban arts investment, including the promotion of "cultural quarters," as designated zones of intensive cultural consumption or production (Bianchini et al. 1988). Confirming the suspicions of some skeptics, who had tended to locate cultural policies on the "soft," populist fringe of the GLC's *industrial* strategy, cultural economy discourse and practice had, by the 1990s, begun to dissipate into conventional urban development agendas (see Bianchini and Parkinson 1993; Mommaas 2004). By the time that the (New) Labour Party formed a national government under Tony Blair in 1997, a reimagined and expansive creative *economy* had been constructed around the vestigial traces of earlier cultural industries efforts, spreading to encompass the high-tech sector and many of the professions, as a funky synonym for the "new economy" (see DCMS 1998; Smith 1998; cf. Frith 2003). A trajectory was duly established in which the self-managing, creative entrepreneur could now be celebrated as an aspirational model for lumpen classes and lagging regions, as creativity became a byword for atomized forms of innovation and 24/7 productivity, and as policymaking began to defer to, instead of critique, the "moral prestige of the 'creative artist'" (Garnham 2005, 26; Peck 2005). This amounted to, in Hesmondhalgh's (2007, 145) words, "the very opposite of the original GLC vision." In this way, the cultural industries projects of the 1980s, which had started with a radical, antiestablishment bang, ended with a conformist, utilitarian whimper. This was a sign, however, of things to come.

Culture as Class Distinction: Neoliberal Urbanism and the Creativity Fix

In the millennial (re)discovery of creativity as an urban asset, little is made of the municipal–socialist legacy, which in a sense really belonged to a different (policy) world. In place of the GLC's inventive policy graft, working knowingly against the grain of both the market and conservative institutions, contemporary creativity policies are presented as feel good, promarket interventions. Gone is the emphasis on democratizing culture economies in favor of marginalized social groups; creativity is now sold as an urban growth strategy, modeled on the achievements and lifestyles of a cosmopolitan elite. This is not, however, simply a matter of the degradation of a once progressive field of policy innovation. The allure, reach, and purchase of Floridian creativity policies arguably reveals more about the political–economic *environment* of urban policymaking in these entrepreneurial, neoliberalizing times (Peck 2009). In this respect, it is not only necessary to pay attention to issues of policy discourse and design, policy talk and technology, but also to consider the expansive *policy ecologies* within which such policies-of-choice are embedded and that enable their conspicuous mobility.

In stylized terms, Figure 3.1 draws a series of distinctions between the policy ecologies associated with municipal socialism and creative urbanism. The ontology of the cultural political economy, as a policymaking imaginary, is dramatically different: the political demonstration projects developed by the GLC were orthogonally positioned *against* a restructuring metropolitan economy, seeking to recognize and empower marginalized cultures and cultural workers; the creative cities script, in contrast, is predicated on a hypercompetitive urban order, lionizing winning sites and subjects, and sanctioning the regressive redistribution of public funds, according to a trickle-down logic, in favor of a deserving overclass. These interventions are embedded in different policy spaces too: if the GLC's strategy was developed by local activist–craftworkers as a disruptive policy technology, *pushed* as part of a transformative political–economic project, creative urbanism is brokered by peripatetic consultant–entrepreneurs, as a means of making over tired growth strategies, and *pulled* by the still unsatisfied demand of urban managers and business-oriented development cadres. Ironically, the demand for creativity policies often originates from real, material need—even if the placebos themselves are tragically inappropriate for these circumstances. Deindustrializing Detroit, in contrast to deindustrializing London a quarter century ago, can be regarded as a critical case in this context. Here, while manifested unfit for the purpose of urban renewal, the tropes and techniques of creativity have nevertheless been sucked into what amounts to a policy vacuum, in a state that had been enduring its own crises of recession and restructuring even before the economic crisis of 2008 hit.

Michigan's Department of Labor and Economic Growth certainly had its hands full during the Bush years. The state's economic fortunes had been tracking those of its principal industry, automobile manufacture, resulting in the highest rates of unemployment and outmigration in the

		Creative cities … municipal socialism	*Cultural industries … neoliberal urbanism*
Reworked policy spaces	*Goal*	Diversity: enrichment of culture	Growth: enrichment through culture
	Impetus	Political demonstration project	Competitive emulation opportunity
	Register	Socialist practice	Market governance
	Restructuring	For labor	For creatives
	Economy	Orthogonal: in and against the market	Complementary: in and for the market
	Redistribution	Positive: social justice	Negative: creatives first
	Ethos	Confrontation	Cool
	Politics	Oppositional	Confirmist
	Subjects	Marginalized producers	Cultural economic elites
	Claim	Participatory engagement	Just desserts
Cultural political economies	*Authorship*	Activist–craftworkers	Consultant–entrepreneurs
	Analytic	Materialist–radical	Idealist–rational
	Locus	Local state	Development regime
	Carrier	Local political activists	Creative growth coalitions
	Technology	Disruptive	Enabling
	Mobility	Innovation/inspiration	Imitation/replication
	Impetus	Political push	Policy pull
	Currency	Bold practice	Best practice
	Repertoire	Experimental–incremental	Reiterative–recurrent
	Coordination	Weak: collaborative learning	Absent: competitive leapfrogging
	Diffusion	Slow–organic	Fast–synthetic

Figure 3.1. Contrasting policy ecologies

nation, rolling fiscal crises across all levels of government, and a multi-year "single-state recession" that seemed unending. The industrial belt of southeast Michigan, which includes Detroit, had shed more than half of its manufacturing jobs since 2000 (Vlasic and Bunkley 2009). For some time, the business community had been demanding action, loudly *naming* the crisis, and insisting on the need to confront the twin challenge of "right-sizing our state and our automotive companies" (Crain 2006, 14). Structural economic problems, by definition, would not be amenable to piecemeal remedies. A Brookings Institution study reported that Michigan had a larger share of "economically weak" cities than any state in the nation, with Detroit, East Lansing, Flint, Kalamazoo, and Saginaw crowding the bottom-of-the-league table (Haglund 2007). But on Valentine's Day 2008, there was at last some good news to report. Governor Jennifer Granholm announced in a Department of Labor press release that Detroit would, in the fall, host the second international Creative Cities Summit: "Michigan's greatest economic successes have always been tied to the creative and productive power of our cities," the governor observed. "We are looking forward to hosting leading thinkers and practitioners from around the world to discuss the latest in creative community development" (Granholm 2008).

The summit's location was anything but accidental. Michigan had gained national attention for its Cool Cities program, which had siphoned $5 million from an overburdened state budget to support scores of neighborhood-scale initiatives for "creative" development, modeled on Richard Florida's increasingly ubiquitous development template (Michigan DLEG 2005). In the hoopla around the program's launch, the governor had posed in sunglasses, declaring that the state's future looked "so hip you will have to wear shades" (quoted in Fischer 2006, C1). The fuse had earlier been lit by the self-styled urban guru, who, in a succession of conferences, events, and consultations, had been egging on local policy, business, and cultural elites to seize the new creative day. Apparently, "optimism reigned" at Florida's first Detroit seminar, where the city's poor performance on the presenter's own creativity scorecard was the subject of near-Orwellian spin—"Detroit . . . has greater potential for improvement" than any other U.S. city, Florida insisted—audience members reporting that they, too, could "feel it . . . there is something that is going to happen" (quoted in *Crain's Detroit Business* 2004, 38).

What happened, in fact, was that steps were quickly taken to establish a creative growth coalition in the city, with the support of all the mainstream economic development players, such as the Detroit Regional Chamber of Commerce, the Detroit Economic Growth Corporation, the city of Detroit, and Detroit Renaissance, together with a coterie of cultural entrepreneurs, public relations and marketing interests, and consultants. Predictably, one of the first moves involved the establishment of Create-Detroit in 2003 (tagline: "bringing the creative class together"), with a remit to energize talent attraction and promotion efforts, mainly through event and Web-based initiatives, social networking, and lobbying for various forms of (artsy or funky) downtown investment.[2] CreateDetroit's first move was to book one of Florida's regional transformation workshops to secure the city's position as a "magnet for new economy talent" (Erickson 2003). The flashy workshop evidently succeeded in propagating the creative buzz, while securing the legitimacy of a rebranded, now-*creative* growth coalition (Peck 2009). The Detroit Regional Chamber of Commerce did its part by establishing Fusion, a networking group for young and "young-thinking" professionals and entrepreneurs, the activities of which included public displays of "fun," schmoozing events styled as *Fusion After 5,* and occasional professional development seminars. In its efforts "to make the Detroit region more attractive to young talent and businesses," Fusion acknowledges that it is working against the tide, because the metro area has been identified as one of the top five "Greatest Brain Draining" regions in the country, "which means the city is losing young educated professionals in droves" (http://www.fusiondetroit.org).

Detroit Renaissance, the region's leading business-led economic development organization, commissioned the consulting firm New Economy Strategies from Washington, D.C., to revamp the metropolitan region's vision and strategy—in the spirit (and in the language) of creative growth. An initial benchmarking exercise found Detroit languishing outside the top forty cities on most key economic development indicators, including the creative cities index, recommending that midtable cities such as "Nashville and Atlanta may have lessons to offer" (NES 2006a, 6). The *Road to Renaissance* plan, launched in 2006, candidly documented a series of "challenges to innovation and competitiveness," including the region's legacy of fraught race relations; a "geopolitical divide" between the city and suburbs; its leadership deficits; its dependency on the "procurement economy" spawned by the automotive supply chain, with stifling effects on entrepreneurialism and new-firm formation; and a pervasive "entitlement mindset," reflecting the "impact of the factory town, the entrenched career pathway, the union relationships, and ultimately the false belief that jobs will be available from one generation to the next" (NES 2006b, 7). The report's authors bluntly stated that "if we are going to fix the region, we must change the culture," which was neatly transposed into an inspirational call for creative mobilization: "If we are to build our region, we *will* create a culture" (NES 2006b, 6, emphasis added).

The report's recommended strategic priorities stood on two legs: first, repositioning the automobile cluster as a global center for mobility and logistics, and second, growing the city's creative economy, while securing the talent base. Resting on the contention that the "creative community is a viable economic engine for the region," the creativity strategy identified undervalued "creative assets" in the visual arts, music, and design; emphasized the need for "unified branding," along with a redoubling of efforts to stimulate cultural tourism and convention trade; offered support for the city of Detroit's recent proposal to establish a cultural district; called for the formation of "artistic incubators" and a creative cluster association; and recommended Austin, Texas, Orlando, Florida, and Providence, Rhode Island, as sources of best-practice inspiration (NES 2006b, 17–19). It made, in other words, all the (now) usual moves.

"We aren't the only region to recognize the value of attracting a creative economy," Detroit Renaissance (2008, 23) rather superfluously confessed. "Many communities, regions, states and countries are [now] focusing on creative sector industries as an economic growth engine." Apparently unselfconscious about the evident lack of originality, but at the same time fostering the impression that these are "policies that work," creativity projects in Philadelphia, Maryland, Ireland, Ontario, Providence, New York, and London were among the "many, many more programs [that Detroit] looked at, studied, learned from, stole from" (25). These borrowed ideas took on a more concrete form with the establishment of a plan for a 3.5-mile "creative corridor" along historic Woodward Avenue, combining a "framework for locating and prioritizing real-estate development," peppered with the usual new urbanist design motifs of walkability, mixed use, and high density (Gensler and KBA 2008, 6). The accompanying creativity strategy, a Motown variant of an intensively replicated package (Lovink and Rossiter 2007a), includes a creative business accelerator, poised to "turn Detroit into a magnet for jobs" (Ilitch 2008, 13A); an expansion of the College for Creative Studies (previously, the College of Art and Design); a creative economy investment fund (contributions welcomed); another rebranding and marketing program; and a new Web portal (http://www.detroitmakeithere) with a mission to "inform, empower and unite the creative community."

Similar sentiments pervaded the Creative Cities Summit 2.0, hosted in Detroit in October 2008. The event attracted around one thousand delegates, twice the number that had attended the first meeting, in Tampa Bay, Florida, in 2004—after which its founder, Peter Kageyama, styled

as a coproducer of Summit 2.0, had effectively franchised the initiative.
Between sessions on retaining talent (sponsored by the Detroit Regional
Chamber) and building creative community assets (sponsored by a local
staffing company), delegates were provided with plenty of opportunities for creative networking. The most intrepid were invited to explore
"Detroit after dark," hobnobbing with entrepreneurs of the city's night-
time economy, and partaking of "martinis or beer, jazz or house music,"
according to taste, before being returned safely to the hotel by 10 p.m.
The highlight of the conference was an überpanel of three of the creativity movement's most energetic proselytizers—Richard Florida, Charles
Landry, and John Howkins—who were credited with the concepts of the
creative class, the creative city, and the creative economy, respectively.
In a wry gesture toward Detroit's old economy, publicly teetering on the
verge of bankruptcy, the session was billed as an audience with the Creative Big Three. The smattering of graduate students who were listening
to this master class would even be able to list the event on their tran-
scripts, after the American Planning Association opted to recognize the
conference as equivalent to 28.5 hours of continuing-education course
credit (*PR Newswire* 2008a), reaffirming the intellectual legitimacy of
the creativity thesis.

   With a characteristic mix of utilitarian economics and new-age liber-
tarianism, the Big Three panel sought to distill the essence of creativity:[3]

> FLORIDA: The association between the creative class, or human capital,
> and rates of innovation and economic growth is just *true*. It's a fact.
> Whether we like it or not, it's an absolute *fact* of economic growth
> and development.
> HOWKINS: The opportunity that we have . . . is to reassert the indi-
> vidual as the main actor and the sole purpose of the society and the
> economic system.
> FLORIDA: The solution to [the] dilemma, of how you balance the ego-
> istic, utility-maximizing, crass, ambitious, individual with the group
> is the neighborhood, it's the community, it's the city. It's not the com-
> pany, it's not the political party . . . We need to be smart, and we need
> to think in those terms.

Notwithstanding such insights, there may have been more to learn outside
the conference than within. America's poorest big city, Detroit has been
hemorrhaging jobs, businesses, and residents for years, while more than
one-third of its population and half its children live under the poverty
line (Wilkinson and Nichols 2008). Median incomes, falling across the
country since 2000, had been plummeting at more than three times the

national average in Detroit (Montemurri et al. 2005). As the financial crisis took hold, "subprime city" would become the mortgage default capital of the country (Ziener 2007). The travails of the Big Three automakers, meanwhile, only seemed to confirm the status of "Detroit" as a synonym for industrial redundancy and corporate failure. Any sharp-eyed planning student, out on the streets during a conference break, could see that this was a "challenging" environment for creativity strategies, too.

Conventionally, the "success" of such strategies is measured in terms of conspicuous forms of (appropriately fashionable) consumption and flamboyant regeneration, manifest at street level. In Detroit, there have certainly been attempts to cater, often quite literally, to the creative class, but according to one local assessment, many of the "loft projects, bars, restaurants and small retailers [that] have tried to base their livelihoods on this emerging group have ended up failing" (Aguilar 2008, C1). The occasional hipster haven, like Cafe D'Mongo's Speakeasy—which prides itself on its location "on a deserted section of Griswold"—has been able to hold on, but in line with the faint pulse of the urban economy only opens one evening per week. For all the concerted efforts of local entrepreneurs and agencies, there remains an uneasy feeling that the talent flows are running in the wrong direction. When the marketing organization Model D[4] dispatched a staff member across the Detroit River to interview Richard Florida's chief statistician, Kevin Stolarick, in Windsor, Ontario, the question was bravely posed: *Could a city really die?* "The scary thing," Stolarick disclosed, with a gesture toward Detroit, "is if something like that is going to happen, it's going to happen the other side of that river" (quoted in Parris 2008, 2). His boss, at least when outside Detroit, has a habit of illustrating the brutal logic of creative class mobility in more glib terms: Jack White abandoned Detroit because the "scene had become too negative and confining," and the locals had become "jealous of The White Stripes' success" (Florida 2008, 25).

Back at the summit, some of the more seasoned observers were experiencing flashes of déjà vu, in this, "the latest effort by established institutions to transform [Detroit's] image from ruined rustbelt to hip hangout" (Aguilar 2008, C1). The eager delegation from Model D was on hand, however, to record "the smart ideas." Their decanted lessons, in fact, represented little more than a credulous list of aspirational homilies and neoliberal platitudes.[5] Something, they recorded, should be done about the fact that Detroit is "a city with a lot of rules," although they were quick to caution: "[W]e're not suggesting creating a lawless city, just one where the rules are modern and make sense and come with a delicious coating of ingenuity." It was time, Model D's creatives suggested, to stop worrying about cuts to public education (as the city embarked on an unprecedented program of layoffs and school closures), and instead seek out "local creatives and even tech companies willing to take on pro-bono work"; after all, "teachers are everywhere." Inevitably, more reasons were discovered to "brush up on marketing 101," responding to the apparently pressing need to position the Detroit Creative Corridor within "the larger D Brand," an increasingly splashy initiative that now has its own annual summit.[6] Cities, the new thinking went, had to learn to *play* like businesses, "municipalities as head hunters, Mayors as talent scouts, City Councils as fun-makers." Cities cannot create cool, however, but they are able to *nurture* it, for example, by booking local bands to play at the airport, another idea from Austin that "Detroit could steal" (Model D Staff 2008).

Detroit's business establishment seems to have had little trouble embracing the creativity credo, while retaining its long-standing positions on public sector austerity and market discipline. Less starchy board members of Detroit Renaissance seem to enjoy the buzz generated by the idea of creative regeneration, and the light relief it must provide from the usual grind of business-retention efforts. But this bastion of the corporate establishment continues to issue shrill pronouncements of impending financial Armageddon—its "modest proposals" for the restoration of business confidence, including a rolling program of government budget cuts, elimination of the Michigan Business Tax, and a $100 million clawback from the Department of Corrections (Brandon and Rothwell 2008). *Crain's Detroit Business,* fresh from the announcement of its sponsorship of Creative Cities Summit 2.0, as an opportunity to "showcase our creative assets [and] connect with creative thought leaders from around the world" (Michelle Darwish, *Crain's Detroit Business,* quoted in *PR Newswire* 2008b) was soon back to the old-style business editorializing:

> Today's crisis puts a harsh spotlight on Michigan's need for dramatic and rapid cost reductions in government . . . eliminating the Michigan Business Tax [and] overhauling public employee and retiree benefits and pensions . . . Michigan [must no longer stand] for status quo, for protecting public employees at the risk of the public good and for business practices dictated by labor agreements that defy the reality of globalization . . . Now, more than ever, the status quo must go. (*Crain's Detroit Business* 2008, 6)

While lobbying for investment in creative real estate, the Detroit business community's other new idea turned out to be an(other) image overhaul, this one with a marketing budget of $1.2 million. *DNews, as the initiative was called,* would be trained on the strategic target of "improving where the Detroit region ranks on a plethora of national surveys about livability and economic viability" (Walsh 2008, 1F). It was with the same circular and self-referential reasoning that Detroit Renaissance had pledged to respond to the city's off-the-charts ranking on indices of creativity, talent retention, and quality of life, by pouring resources into the creative corridor, and into a newly "unified" rebranding effort, with a view to securing, rather than material change itself, "improved rankings in independent cultural assessment surveys" (NES 2008b, 19; 2008a). In such a way, urban rankings have not only become an important register through which intensified interurban competition is played out, subtly canalizing the means and ends of urban development (McCann 2004; Lovink and Rossiter 2007b), but also play a crassly instrumental role in rendering (supposedly) calculable the nebulous objectives of creativity makeovers (chapter 4).

The vacuity of these efforts was further revealed when Karen Gagnon, director of Michigan's Cool Cities program, appeared on Carol Coletta's *Smart City* public radio show, to promote the Motown summit.[7] Responding to a series of softball questions from the show's host, a creative cities consultant, Gagnon conceded that her state had not done "enough of a good job of preparing for" the apparently imminent demise of the local automobile industry, but if there was a silver lining in the crisis, it was that it had generated a newfound "openness to some new ideas . . . related to creative place-making [and] the new-economy paradigm [which] the majority of our cities [have] really embraced." Detroit needed a *symbolic* makeover, she claimed, trading in the old image, as the epitome of industrial urbanism, for a more allegorical role in the spatial division of labor, as "a metaphor for . . . changing to a new-economy paradigm, meaning not all dependent on the old ways . . . Everything is changing." This feat could apparently be accomplished by prepping the creative supply side, establishing "the right environment, to allow people to be creative." The creativity script's prefabricated tropes—appealing, as they do, to new age urbanism, parochial pride, and can-do optimism—seemed to be just about adequate for selling on this policy position, that is, until the host posed an obvious, if rather tricky, question:

> COLETTA: Does the idea of a creative city have any meaning to the people who make automobiles and auto parts?
>
> GAGNON: I have not sat down [laughs nervously] with the executives of some of the Big Three or Big Four . . . It's hard for me, you know, to actually say yes or no.
>
> COLETTA: But your governor certainly has, Governor Granholm certainly has . . . and you have run her Cool Cities initiative I think since

the very beginning . . . I'm just curious, when she sits down with automakers, have you ever heard her say if this has any resonance for them? GAGNON: I know that she has said a few things here and there. There is definitely some resonance. I would like to see more. I think there is always room for more resonance with the Big Three companies.[8]

In light of the predictable shortage of resonance, Coletta really should have known better, for she was hardly new to this territory. As the eponymous director of the consulting firm, Coletta & Company (now Smart City Consulting), she was one of a handful of key players in the simultaneous mobilization, during 2002–3, of the creative cities project as an elite social movement and as a new age consultancy product. Her hometown of Memphis, Tennessee, had ranked dead last on Florida's top fifty chart of creative cities, several places behind other allegedly buzz-free zones, like Detroit, Buffalo, and Las Vegas (Florida 2002, table 13.3). Rather than alienating or infuriating the stunted hipster class of the Home of the Blues, it spurred them to action. Coletta rushed to convene the Memphis Manifesto Summit, in association with Richard Florida and the city's regular lineup of economic development players, while also setting in motion what would become the Memphis Talent Magnet Project (see MTMP and Coletta & Company 2003). Within months, countless other cities were following suit. For example, the Tampa Bay Partnership, the Sarasota County Commission for Economic Development, and the Florida High Tech Corridor quickly got together to commission a copycat "young and restless" study for Tampa Bay, which had also felt the competitive pinch after its midtable ranking in *The Rise of the Creative Class* (Impresa and Coletta & Company 2004).[9] This group, too, had formed a creative "front" organization, with the mandatory compressed glyphics, just in case anyone should miss the point: CreativeTampaBay predictably indulges in entirely typical forms of viral rebranding, modish marketing, and cocktail-fueled opportunities for "casual conversations and catalytic networking," together with self-congratulatory events, such as an annual homage to the region's forty more creative movers and shakers.

This organization also has its origins in a Richard Florida event, in April 2003, after which four local creatives followed the guru to Memphis to become signatories of the Memphis Manifesto (see Peck 2005). On their return, in addition to establishing their own "grassroots organization," following the formula recommended for other creative wannabes, the Tampa 4 persuaded the local council to appoint a creative industries manager and set about to organize the inaugural Creative Cities Summit.

The impresario of this latter venture was Peter Kageyama, self-styled social entrepreneur, marketing consultant, and partner of one of the Tampa 4. As president of CreativeTampaBay, Kageyama would develop an increasingly peripatetic lifestyle, peddling the creativity credo everywhere from Perth, Australia, to Huddersfield, England. Understandably, the market for recycled creativity policies is much wider at the bottom of the urban hierarchy, where the creative talent pool is supposedly thinner (according to the "theory") and where there is often more than a whiff of desperation in efforts to climb aboard the cool caravan.

The medicine show's arrival in Detroit was therefore effectively preordained. "Detroit is interesting as a metaphor," Kageyama opined at the second summit, for what the industrial city had (once) been able to achieve, though clearly it had "now fallen on challenging times" (quoted in Ankeny 2008, M15). As the region's auto industry tumbled into receivership, the times would become even more challenging. A $100 million "New Economy Initiative," launched in March 2009, offers grants for innovative ideas that might turn the economic tide, underlining the imperative for the region to be "warm and welcoming to creative, talented people of all backgrounds" and publicly committing to a strict, measurement-by-metrics approach to impact analysis (NEI 2009, 3). Meanwhile, the city of Detroit, which labors under the largest municipal debt in the country (at 20 percent of general funds), contemplates whether to rededicate its swathes of abandoned land to theater projects, windfarming, or some other form of "creative downsizing" (*Economist* 2009, 34).

Conclusion: Policy Cultures, Policy Ecologies

The two moments of "creative" urban policymaking contrasted here, while sharing the broad goal of leveraging "culture" in the service of urban development, are associated with markedly different rubrics, rationalities, and ramifications. During the early 1980s, Labour-controlled councils like the GLC represented localized bulwarks against Thatcherism at the national scale. Their advocacy of cultural industries policies occurred in the context of an antimonetarist posture, drawing on the expertise of activist policymakers and favoring provocative, politicized, and practice-oriented forms of interventionism. The municipal–socialist project relied heavily on politically targeted, strategic interventions in metropolitan economies, complemented with a rudimentary system of extralocal policy learning between politically aligned local authorities. The contrast with today's globalizing fast-policy "market" for creativity fixes is a stark one. While this is (still) mostly fueled by public funds, it is prefiguratively animated by a distended complex of academic entrepreneurs, consultants, conferences, blogs, and Web sites, morphing into an imitative network of creative growth coalitions at the urban scale (chapter 4). Whereas the GLC's policies were negotiated between county hall "craftworkers" and a series of unruly external political constituencies, their contemporary successors are more often sold by consultants, promoted through marketing networks, anointed by corporate–establishment agencies, and affirmed at "summits."

In truth, unambiguously positive policy outcomes have been elusive in both of these episodes, but the velocity and scope of contemporary creativity fixes far exceeds their 1980s predecessors. The speed and reach of contemporary forms of "creative policy transfer," and the sheer promiscuity of policy practice in this field, certainly cannot be explained in terms of the intrinsic effectiveness of the policy measures themselves. Rather, they spread because they conform, not because they "work." Nouveau creativity policies are carriers for fiscally undemanding forms of market-complementing interventionism; they are minimally disruptive of extant power structures and established interests, and they "accessorize" neoliberal urbanism in a manner befitting prevailing cultural tropes of competitive cosmopolitanism. The rationale of the municipal–socialist project, in contrast, was to confront and counter the neoliberal policy orthodoxy of the early 1980s. The GLC was actively *producing* an alternative, countercultural policy project; Detroit has been *consuming* a transnational policy fix.

The homegrown cultural industries initiatives of the GLC and the creative economy packages sold to cities like Detroit, therefore, should not be distinguished merely in the language of "policy choices." Fundamentally, they derive from different ontologies of the creative economy, different constructions of creative work, and different rationalities of intervention. The municipal socialist strategy was concerned with extending the registers of the economic, in part by valuing (hitherto marginalized) forms of cultural distribution and production. But while the GLC sought to uncover and revalue cultural work in the cracks, as it were, of the formal economy, the Floridian creativity thesis promotes (innate) "talent," along with associated creative "assets," as the supply-side drivers of positive (indeed, ascendant) forms of urban economic growth. The GLC's policies were designed to broaden the field of cultural recognition, while opening up access to livelihood strategies, for marginalized social groups and would-be cultural workers. In contrast, today's creativity script is an aspirational narrative, celebrating the business achievements and lifestyle choices of creative entrepreneurs, their self-indulgent forms of overwork, expressive play, and conspicuous consumption. Whereas the cultural industries project of the early 1980s sought to challenge existing distributions of cultural value and cultural labor, formulating progressive policies for cities that flew in the face of political–economic orthodoxy, the creativity script effectively affirms and exaggerates them, thereby relegitimizing and reinforcing both market-confirming practices and regressive social redistributions within the city.

The GLC's interventions were knowingly strategic and interstitial, framed in the context of a materialist understanding of (metropolitan) economic restructuring and a (generally) shrewd reading of the landscape of political opportunity in the city. In contrast, the imported creative city interventions of today are framed by an entirely different calculus, deferring to a new *logic* of knowledge-driven growth with a package of minimally disruptive "add-ons" to extant urban economic development programs. While the two strategies might both be considered part of the same urban policy genealogy, Hesmondhalgh (2007) surely has a point in characterizing today's creativity cults as practically the *opposite* of their municipal–socialist predecessors.

While the GLC used culture as a tool of political mobilization within the city, the creativity projects of today are more about mobilizing a new generation of culturally tinged growth coalitions across cities, reconstituting external competitive threats in novel terms, and defining new aspirations-cum-responsibilities for a broadened network of urban policy protagonists. (What, after all, is a "summit," if not a gathering of elite decision makers, focused on the search for a practical solution to an extant crisis?) The "quicksilver adoption" of urban creativity strategies, which has been described by Ross (2006, 27) and others, must be understood in light of the following conditions: first, their productive conformity with urban growth–machine interests; second, their broad compatibility with selective, supply-side policymaking in image, labor, and housing markets; and third, their fundamental complementarity with the (still-dominant, but always evolving) neoliberal development ethos. Working "in and against the market," GLC-style, has apparently been superseded by working in *and for* the market. This has been more than an ideological realignment. It has been accompanied by, and realized through, a transformed apparatus of urban policy formation and circulation. The velocity and promiscuity of creative cities policies is hardly a product of their efficacy, the returns to which remain elusive (almost to the point of invisibility), though they surely do testify to the scope and connectivity of fast urban policy networks. The creativity script arrives with ready-made rationales, remits, and rhetorics, functioning as a carrier of trendy policy norms, yet requiring only minimal adaptation to local circumstances. This forward- and upward-looking script addresses, and readily enrolls, emergent local complexes of knowing creative subjects, assigning lucrative new roles to consultant–entrepreneurs, while reaching deep into established urban development communities. As a result, creative growth coalitions— key nodes in these fast-policy networks—seem to be metastasizing all over the place, where the hunger for rebadged policy formulations seems only matched by a pervasive, if not existential, sense of competitive anxiety. The policy craftworkers at the GLC twenty-five years ago would have recognized (and acknowledged) the nature of these competitive threats, though they would surely have been bemused and perplexed by the nature of the contemporary response.

Notes

Thanks to Amy Cervenan, Allan Cochrane, Eugene McCann, Neil McInroy, Lyn Stacey, and Kevin Ward. Responsibility for the interpretations here is, however, mine.

1. Michael Ward, chair of the Industry and Employment Committee at the GLC, would later become the longtime director of CLES.

2. A local marketing company decided that it could spare one of its young executives, Eric Cedo, to serve as CreateDetroit's founding CEO, although he would only stay for a year before moving on to open his own consulting operation, Brain-Gain Marketing.

3. Richard Florida and John Howkins, Creative Cities Summit 2.0, transcribed from http://www.youtube.com/watch?v=laJqclydrkI.

4. Model D, which hosts a weekly e-zine and Web site, was formed with the intention of "creat[ing] a new narrative for Detroit—telling the stories of its development, creative people and businesses, vibrant neighborhoods and cool places to live, eat, shop, work and play."

5. Following quotes from Model D Staff (2008, 2–4).

6. "The D Brand Summit is a groundbreaking, one-day annual event that brings professionals together to share cutting-edge brand-building skills, resources and insights that can make a real difference in the success of your organization— and the destiny of the D." These efforts are deemed necessary because "competition for new visitors and bright talent is heating up. What strategies and stories are other cities and companies using to stake their claim? How can the D hold its own and even gain new ground? How can your company capture the best talent and avoid the brain drain? What kind of role do you as a marketer stand to play in helping your region to get ahead—and in helping your organization to get ahead with it?" (http://dbrandsummit.com)

7. The following quotes are transcribed from *Smart City*, October 2, 2008, accessed at http://www.smartcityradio.com.

8. Transcribed from *Smart City*, October 2, 2008, accessed at http://www .smartcityradio.com.

9. Coletta and her collaborators later sold young and restless studies, effectively straight off the shelf, to creative growth coalitions in Providence, Rhode Island, Philadelphia, Pennsylvania, Richmond, Virginia, and Portland, Oregon, with identical covers and identical tables of contents (see http://www.restlessyoung.com).

References

Abrahamson, E. 1996. "Management Fashion." *Academy of Management Review* 21:254–85.

Aguilar, L. 2008. "Hip or Miss? Detroit Attempts to Increase Cool Factor, Appeal to Creative Crowd." *Detroit News,* February 5, C1–C2.

Alcock, P., and P. Lee. 1981. "The Socialist Republic of South Yorkshire?" *Critical Social Policy* 1:72–93.

Alcock, P., A. Cochrane, and P. Lee. 1984. "A Parable of How Things Might Be Done Differently." *Critical Social Policy* 9:69–87.

Ankeny, R. 2008. "Creative Cities' Meeting Reaches beyond Arts, Culture." *Crain's Detroit Business.* June 2, M15.

Bailey, J. 1989. *Marketing Cities in the 1980s and Beyond.* Washington, D.C.: American Economic Development Council.

Barnett, A. 1985. "Whistling in the Wind?" *New Socialist*, January, 17–20.

Bianchini, F. 1987. "Cultural Policy and Changes in Urban Political Culture: The 'Postmodern Response' of the Left in Rome (1976–85) and London (1981–86)." Paper presented at European Consortium for Political Research workshop, Amsterdam, Netherlands, April 10–15.

———. 1989. "Cultural Policy and Urban Social Movements: The Response of the 'New Left' in Rome (1976–1985) and London (1981–1986)." In *Leisure and Urban Processes*, edited by P. Bramham, I. Henry, H. Mommaas, and H. van der Poel, 18–46. London: Routledge.

Bianchini, F., M. Fisher, J. Montgomery, and K. Worpole. 1988. *City Centres, City Cultures.* Manchester, UK: Centre for Local Economic Strategies.

Bianchini, F., and M. Parkinson, eds. 1993. *Cultural Policy and Urban Regeneration.* Manchester, UK: Manchester University Press.

Blunkett, D., and G. Green. 1983. *Rebuilding from the Bottom: The Sheffield Experience.* Tract 491. London: Fabian Society.

Brandon, D., and D. Rothwell. 2008. "Bold State Fiscal Reform Critical for Business." *Detroit Free Press*, December 1, 12A.

Campbell, M., M. Hardy, N. Healey, R. Stead, and J. Sutherland. 1987. *Economic Sense: Local Jobs Plans—A National Perspective*. Manchester, UK: Centre for Local Economic Strategies.

Centre for Local Economic Strategies (CLES). 1986. *Annual Report*. Manchester, UK: CLES.

Cochrane, A. 1986a. "Local Employment Initiatives: Towards a New Municipal Socialism? In *The Contemporary British City*, edited by P. Lawless and C. Raban, 144–62. London: Harper and Row.

———. 1986b. "What's in a Strategy? The London Industrial Strategy and Municipal Socialism." *Capital and Class* 28:187–93.

———. 1988. "In and Against the Market? The Development of Socialist Economic Strategies in Britain, 1981–1986." *Policy and Politics* 16:159–68.

Crain, K. 2006. "We Really Are in a Single-State Recession." *Crain's Detroit Business*, November 6, 14.

Crain's Detroit Business. 2004. "Rumblings." March 8, 38.

———. 2008. "Legislators Must Cut Cost of Government." October 6, 6.

Davenport, E., S. Maddock, D. Miller, and D. Mort. 1987. *Against the Odds: Local Job Generation*. Manchester, UK: Centre for Local Economic Strategies.

Department of Culture, Media and Sport (DCMS). 1998. *Creative Industries Mapping Document*. London: DCMS.

Detroit Renaissance. 2008. *Growing Detroit's Creative Economy: Road to Renaissance Initiative*. Detroit, Mich.: Detroit Renaissance.

Economist. 2009. "Baptism by Five-Alarm Fire." May 30, 34.

Eisenschitz, A., and D. North. 1986. "The London Industrial Strategy: Socialist Transformation or Modernising Capitalism?" *International Journal of Urban and Regional Research* 10:419–39.

Erickson, H. 2003. "CreateDetroit. Who Needs It? We Do!" *Detroiter*, December, 62.

Fischer, D. 2006. "The Case For/Against Cool Cities." *Ottawa Citizen*, June 18, C1.

Florida, R. 2002. *The Rise of the Creative Class*. New York: Basic Books.

———. 2005. *Cities and the Creative Class*. New York: Routledge.

———. 2008. *Who's Your City?* New York: Basic Books.

Frith, S. 2003. "Mr. Smith Draws a Map." *Critical Quarterly* 41:3–8.

Garnham, N. [1983] 1990. "Public Policy and the Cultural Industries." In *Capitalism and Communication*, edited by N. Garnham, 154–68. London: Sage.

———. 2005. "From Cultural to Creative Industries: An Analysis of the Implications of the 'Creative Industries' Approach to Arts and Media Policy Making in the United Kingdom." *International Journal of Cultural Policy* 11:15–30.

Geddes, M. 1988. "The Capitalist State and the Local Economy: 'Restructuring for Labour' and Beyond." *Capital and Class* 35:85–120.

Gensler and KBA. 2008. *Detroit Renaissance Creative Corridor Development Plan*. New York: Gensler and KBA.

Greater London Council (GLC). 1984. *Cultural Industries Strategy*. London: Industry and Employment Committee, GLC.

———. 1985. *London Industrial Strategy*. London: GLC.

Goodwin, M., and S. Duncan. 1986. "The Local State and Local Economic Policy: Political Mobilisation or Economic Regeneration." *Capital and Class* 27:14–36.

Graham, J. 1992. "Post-Fordism as Politics: The Political Consequences of Narratives on the Left." *Environment and Planning D: Society and Space* 10:393–410.

Granholm, J. 2008. "Governor Granholm Announces Detroit Will Host International Creative Cities Summit in October." *USFed News*, February 14. http://www.michigan.gov/dleg/0,1607,7-154—185600—,00.html.

Green, G. 1987. "The New Municipal Socialism." In *The State or the Market: Politics and Welfare in Contemporary Britain*, edited by M. Loney, 203–21. London: Sage.

Greenhalgh, L., O. Kelly, and K. Worpole. 1992. "Municipal Culture, Arts Policies, and the Cultural Industries." In *Restructuring the Local Economy*, edited by M. Geddes and J. Bennington, 120–36. Harlow: Longman.

Gyford, J. 1985. *The Politics of Local Socialism*. London: Allen and Unwin.

Haglund, R. 2007. "Michigan Cities Must Work on Being Cool." *Grand Rapids Press*, May 23, C3.

Hall, S. 1984. "Face the Future." *New Socialist*, September, 37–39.

Hesmondhalgh, D. 2007. *The Cultural Industries*. London: Sage.

Hosken, A. 2008. *Ken*. London: Arcadia.

Ilitch, C. 2008. "Creative Accelerator Can Turn Detroit into Magnet for Jobs." *Detroit News*, October 15, 13A.

Impresa and Coletta & Company. 2004. *The Young and the Restless: How Tampa Bay Competes for Talent*. Portland, Ore.: Impresa and Coletta.

Jessop, B., K. Bonnett, S. Bromley, and T. Ling. 1988. *Thatcherism*. Cambridge: Polity Press.

Lovink, G., and N. Rossiter, eds. 2007a. *MyCreativity Reader: A Critique of Creative Industries*. Amsterdam, Netherlands: Institute of Network Cultures.

———. 2007b. "Proposals for Creative Research: Introduction to the *MyCreativity Reader*." In *MyCreativity Reader: A Critique of Creative Industries*, edited by G. Lovink and N. Rossiter. Amsterdam, Netherlands: Institute of Network Cultures.

Mackintosh, M., and H. Wainwright. 1987. *A Taste of Power*. London: Verso.

Massey, D. 1997. "A Feminist Critique of Political Economy." *City* 2:156–62.

McCann, E. J. 2004. "'Best places': Interurban Competition, Quality of Life and Popular Media Discourse." *Urban Studies* 41:1909–29.

Memphis Talent Magnet Project (MTMP) and Coletta & Company. 2003. *Technology, Talent, and Tolerance: Attracting the Best and Brightest to Memphis*. Memphis, Tenn.: Coletta.

Michigan Department of Labor and Economic Growth (DLEG). 2005. *Michigan Cool Cities Initiative*. Lansing: State of Michigan.

Model D Staff. 2008. "Smart Ideas for Detroit from the Creative Cities Summit 2.0." *Model D*, October 21. http://www.modeldmedia.com/features/summit 16508.aspx.

Mommaas, H. 2004. "Cultural Quarters and the Post-Industrial City: Towards the Remapping of Urban Cultural Policy." *Urban Studies* 41:507–32.

Montemurri, P., K. Gray, and C. Angel. 2005. "Detroit Tops Nation in Poverty Census." *Detroit Free Press*, August 31, 3B.

Moss, L. 2002. "Sheffield's Cultural Industries Quarter 20 Years On: What Can Be Learned from a Pioneering Example?" *International Journal of Cultural Policy* 8:211–19.

Mulgan, G., and K. Warpole. 1986. *Saturday Night or Sunday Morning? From Arts to Industry*. London: Comedia.

Murray, R. 1983. "Pension Funds and Local Authority Investments." *Capital and Class* 20:89–103.

———. 1987. *Breaking with Bureaucracy*. Manchester, UK: Center for Local Economic Strategies.

New Economy Initiative for Southeast Michigan (NEI). 2009. *Accelerating the Transition of Metro Detroit to an Innovation-Based Economy*. Detroit, Mich.: New Economy Initiative.

New Economy Strategies (NES). 2006a. *Competitive Intelligence Benchmarking Tool: Greater Detroit*. Washington, D.C.: NES.

———. 2006b. *Road to Renaissance: A Collaborative Strategy for Regional Economic Growth*. Washington, D.C.: NES.

Nolan, P., and K. O'Donnell. 1987. "Taming the Market Economy? A Critical Assessment of the GLC's Experiment in Restructuring for Labour." *Cambridge Journal of Economics* 11:251–63.

Parris, T. 2008. "Man behind the Creative Class Stats Shares Ideas on Detroit." *Model D*, December 2. http://www.modeldmedia.com/features/stolarick170.aspx.

Peck, J. 2005. "Struggling with the Creative Class." *International Journal of Urban and Regional Research* 24:740–70.

———. 2009. "The Cult of Urban Creativity." In *Leviathan Undone? Towards a Political Economy of Scale*, edited by R. Keil and R. Mahon, 159–76. Vancouver: University of British Columbia Press.

Peck, J., and N. Theodore. 2010. "Mobilizing Policy: Models, Methods, and Mutations." *Geoforum* 41:169–74.

PR Newswire. 2008a. "American Planning Association to Grant Continuing Education Credit for Attendees at Creative Cities Summit 2.0." October 1. http://www.reuters.com/article/pressRelease/idUS230789+01-Oct-2008+PRN20081001.

———. 2008b. "Crain's Sponsors Creative Cities Summit 2.0." September 25. http://www.reuters.com/article/pressRelease/idUS247171+25-Sep-2008+PRN 20080925.

Quilley, S. 2000. "Manchester First: From Municipal Socialism to the Entrepreneurial City." *International Journal of Urban and Regional Research* 24:601–15.

Ross, A. 2006. "Nice Work If You Can Get It: The Mercurial Career of Creative Industries Policy." *Work Organisation, Labour, and Globalisation* 1:13–30.

Rustin, M. 1986. "Lessons from the London Industrial Strategy." *New Left Review* 155:75–84.

Scott, A. J. 2006. "Creative Cities: Conceptual Issues and Policy Questions." *Journal of Urban Affairs* 28:1–17.

Smith, C. 1998. *Creative Britain*. London: Faber and Faber.

Vlasic, B., and N. Bunkley. 2009. "Life without King Auto." *New York Times*, June 10, B1–B4.

Wainwright, H. 1984. "The Greater London Council's Economic Program." *Working Papers for Policy Alternatives* 1, Ottawa, ON: Canadian Centre for Policy Alternatives.

Walsh, T. 2008. "As Detroit Auto Economy Sputters, a New Image-Boosting Campaign Revs Up." *Detroit Free Press*, November 9, 1F–4F.

Ward, M. 1981. "Job Creation by the Council: Local Government and the Struggle for Full Employment." Pamphlet 78. Nottingham, UK: Institute for Workers Control.

———. 1983. "Labour's Capital Gains: The GLC Experience." *Marxism Today*, December, 24–29.

Wilkinson, M., and D. Nichols. 2008. "Michigan Income Down, Poverty Up." *Detroit News*, August 27, B1.

Williams, R. 1958. *Culture and Society 1780–1950*. New York: Columbia University Press.

Ziener, M. 2007. "Dateline Detroit: Welcome to Subprime City." *The Globalist*, October 17–18. http://www.theglobalist.com/StoryId.aspx?StoryId=6513.

CHAPTER FOUR

Policies in Motion and in Place

The Case of Business Improvement Districts

Kevin Ward

It is unlikely that many of you have heard of Royston in England. If you have, I apologize. It is a small market town approximately fifteen miles southwest of Cambridge and forty miles north of London, with a population of just over fifteen thousand. Not known for much, beyond being where the world's first catalytic converter was invented, it has in recent years experienced a gradual downturn in its economic fortunes. While its retail center retains a mix of independent and multinational outlets, those overseeing its future economic viability have become concerned. Initially, a town center manager was appointed, and a Royston Town Centre Forum was established. This undertook various marketing and promotional activities in an attempt to make the center more competitive. Although it had its successes, those in the public and private sectors were also aware of its limits. A senior manager at Johnson Matthey, a locally based but international-in-reach chemicals company, with operations in more than thirty countries and employing almost nine thousand staff worldwide, together with a mixture of local store owners and representatives of the city and county government, were aware of how other localities facing similar issues were responding. The fate of Royston was compared unfavorably with neighboring Rugby, for example. So began the long process of petitioning for the creation of a Business Improvement District (BID), a public–private partnership in which property and business owners in a defined geographical area vote to make a collective contribution to the governance of their commercial district. Businesses were questioned on their aspirations for the town center, an external consultant was hired to "consult" on the process, and the formal legal procedures were put in place for the vote. On December 2, 2008, Royston became the latest English locality to establish a BID when 39 percent of the 61 percent of all businesses that voted agreed to establish a Royston BID. Royston First, the group of local stakeholders who led the campaign to establish a BID, formed the not-for-profit private company when the Business Improvement District came into existence on April 1, 2009. It will have five years to deliver on its objectives of "transforming the look and feel of the town, reversing the negative image, increasing footfall, spend and dwell time" (Royston First 2008, 2). It will not be doing this in an informational vacuum. As Royston First acknowledges, "There are 76 BIDs operating throughout the UK, so we are able to learn much about what works and what the impact of projects in a BID area are." Add to this a whole plethora of organized events—conferences, master classes, seminars, and workshops—and online resources, most notably through the National BIDs Advisory Service (http://www.ukbids.org)—and it is evident that a vast informational infrastructure is in place to support the rolling out of the BID model across the country.

It is hard to imagine a place more dissimilar to New York than Royston. One might also say the same about Bedford (a small town in Bedfordshire in southeast England), Oldham (a medium-sized town in northwest England), and Winchester (a small town in southwest England). Yet to understand what will take place in Royston over the next five years, it is necessary to turn to New York. More specifically, about how what happened in that city in the 1990s involving Business Improvement Districts was behind the UK government's 2001 introduction of a BID model, which, in turn, shaped the way differently situated and sized cities and towns in England, such as Royston, have strived to revalorize their central economies. During this period in New York, the BID model came to be seen as an important element in the strategy for turning around the previously ailing Manhattan economy. According to one local commentator, "[Business Improvement Districts] . . . cleaned up, almost alone, key areas of Manhattan business districts . . . They . . . contributed to the drop in crime" (Lentz 1998, 4). While the actual origins of the model lay elsewhere (Hoyt 2006; Ward 2006), and other U.S. cities had also introduced Business Improvement Districts with differing degrees of success (Mitchell 2001; Ward 2007a, 2010), it was the model's place in the internationally heralded successful renaissance of New York during the 1990s that alerted practitioners, policymakers, politicians, and other agents of transference to the possibilities of the BID model, setting the context for the recent vote in Royston.

This chapter uses the internationalization of Business Improvement Districts to address the volume's examination of the global circuits of knowledge, the institutions, and the individual actors involved in making contemporary cities into governable spaces. It is in three sections. The first provides a short argument for how best to understand theoretically the way cities are produced through the circuits, networks, and webs that connect them unevenly and in a geographically differentiated manner. This is not about supplanting more traditional and still valuable territorial analyses of "urban" politics. Instead, it highlights the types of agents of transference involved in moving and embedding urban development programs in general, focusing on its most visible channels (chapters 2 and 5). The second section discusses the context for Business Improvement Districts. Then, the chapter moves on to outlining the model's relational and territorial geographies. It presents an analysis that emphasizes both the ways in which the BID model has been rendered mobile and the place-specific contextual factors that shape how the model has been territorially embedded in different places.

Relational and Territorial Geographies

If it were ever enough to account for change in the nature of urban development on the basis of analysis generated solely from within cities and the countries of which they are part, then that time has surely passed. Recent work from across the social sciences has, instead, argued for an approach to the relational and territorial theorization of space (Massey 1993, 1999, 2007; Allen et al. 1998; Allen and Cochrane 2007; Morgan 2007; McCann and Ward 2010). It has highlighted the increasingly open, porous, and interconnected configuration of territorial entities (Massey 2005, 2007). As MacLeod and Jones (2007, 1186) put it, "all contemporary expressions of territory . . . are, to varying degrees, punctuated by and orchestrated through a myriad of trans-territorial networks and relational webs of connectivity." One such consequence for urban studies scholars has been a burgeoning interest in the circuits, networks, and webs in and through which policies and models are transferred (Peck and Theodore 2001; Ward 2006, 2007b; Robinson 2006; Cook 2008; McCann 2008, 2011; McCann and Ward 2010). Eschewing the political science–dominated "policy transfer" literature, which although not without its insights is also not without its limits (Ward 2007a; McCann 2008; McCann and Ward 2010; Prince 2010), this work has sought to uncover how—through what practices, where, when, and by whom—urban policies are produced in a global relational context, are transferred and reproduced from place to place, and are negotiated politically in various territories. For Wacquant (1999, 321), there is a need to "reconstitute, link by link, the long chain of institutions, agents, and discursive supports (advisors' memoranda, commission reports, official missions, parliamentary exchanges, expert panels, scholarly books and popular pamphlets, press conferences, newspaper articles and televisions reports, etc.)." Larner (2003, 510) agrees, arguing for a "more careful tracing of the intellectual, policy, and practitioner networks that underpin the global expansion of neoliberal ideas, and their subsequent manifestation in government policies and programmes." Specifically, it is argued that paying particular attention to how, through "ordinary" and "extra-ordinary" activities, and in and through "globalizing micro-spaces" (Larner and Le Heron 2002, 765), such as conferences, seminars, and workshops would reveal how and why some policies are made mobile while others are not, and what this means for the ongoing sociospatial (re)structuring of cities (chapter 5).

Of course, this interest in the mobility of models should not be taken to suggest that "territory" and "place" are academically redundant concepts. Far from it, and indeed, an important aspect in this expanding intellectual field is acknowledging the importance of the territorial embedding of mobile policies in particular places (Peck and Theodore 2001; McCann 2008, 2011; McCann and Ward 2010). Approaches that emphasize "territory" and "scale" in forging and representing the political coordinates of the current global urban condition continue to have strong intellectual purchase (Brenner 2004; MacLeod and Jones 2007). There is a long ideational lineage to these contributions, particularly around notions of "coalitions," "regimes," and other sorts of territorial alliances (Logan and Molotch 1987; Cox and Mair 1988; Harvey 1989; Stone 1989; Cox 1995; Peck 1995; Lauria 1997; Jonas and Wilson 1999; Ward 2000).

Understanding and explaining the making, unmaking, and remaking of spatial restructuring and transformation demands we identify the geographical asymmetrical power-geometries that underpin them (Massey 1993; Sayer 2004). This necessitates a continued attention to territoriality, as "a micro-world that is experienced and contested as a lived space; as heterogeneity negotiated habitually through struggles over roads and noise, public spaces, siting decisions, neighborhoods and neighbors, housing developments, street life and so on" (Amin 2004, 39). "Many prosaic moments of realpolitik" (MacLeod and Jones 2007, 1185) are then expressed and conducted through territorial, or "turf," politics of dependence and engagement (Cox 1998).

In the example of embedding mobile policies, mobility remains heavily structured and stratified by issues of absolute and relative place-based characteristics (chapter 3). Past developments within and between territories make some mobility more or less likely. As McCann (2011) puts it, "one's embeddedness in particular institutional and political contexts . . . define[s] a constrained set of pathways for action." "The policy development process is in large part a path-dependent one," Peck and Theodore (2001, 430) remind us. The making mobile of models occurs in the context of a "discussion of existing problems, general ideas about dealing with them, and specific proposed solutions," according to Wolman (1992, 34), most of which will be structured by past and existing patterns and processes of development within territories.

Historicizing the Mobility of Policies and the
Disciplining Effects of Comparison

Of course, it is not the case that in the past those involved in designing and planning urban revitalization lived in splendid isolation. They did not (Hall 1989; Wolman 1992; Campbell 2001). There is a reasonably long history to the mobility of urban policies and programs. According to Hoyt (2006, 223), "for hundreds of years, urban policy entrepreneurs—like architects, planners and other experts—have traveled to study other places, make contacts, attended lectures and returned to homelands to report what they had learned." For example, the construction of colonial cities involved the widespread transfer of policies and programs from one colony to another as particular political economic pathways and trajectories were established (Blaut 1993). Within Europe, the period from the early twentieth century saw the establishment of a number of "transboundary" "municipal connections" to the point that

> the network of individual exchanges, visits, writings and their circulation, whether as part of an external structure (a political party, trade union or academic conference) or as an activity within an association in the municipal world itself . . . gradually built up a continuum of experience in, and knowledge of, municipal government. (Saunier 2002, 518)

This is not to argue that nothing has changed in the world of policy transfer (Healey and Upton 2010). It has. In Europe, the 1950s onward saw a change in the nature of the connections between municipalities as welfare states were restructured, while from the 1960s onward, the urban development pathways of the UK and the United States have become increasingly intertwined (Wolman 1992). According to Peck and Theodore (2001, 429), the 1990s were witness to a "substantially narrow form of 'fast policy transfer' between policy elites, based on a truncated and technocratic reading of program effectiveness, coupled with truncated processes of policy formation and evaluation." So it has continued in the 2000s. Behind this changing context for the making mobile of policies and programs has been the growth in the number of translocal urban consultancy organizations and the associated rise of a new global consultocracy, the establishment of new, or expanded, policy networks around think tanks, governmental agencies, professional associations, trade unions, and nongovernmental organizations, and the expansion in international conferencing and "policy tourism" (Peck 2003; Ward 2006, 2007a; McCann 2008; McCann and Ward 2010). The emergence globally of the redevelopment "professional"—part economist, part engineer, part planner, part marketing executive—has been an important outcome of, and contributing factor toward, this rapid-fire, no-questions-asked movement of policies and programs, of which the BID model is illustrative.

The other side to this increased making mobile of models is the extra emphasis on comparison and its calculation by agents of transference (Ward 2008, 2010; Robinson 2008; McFarlane 2011). According to Larner and Le Heron (2002, 417), in recent years, "the global has become more knowable by placing the experiences and performances of others into quantitatively and qualitatively encoded proximity." Cities are the center of this new knowable global context. Through the process of translation—the means through which governance is performed over a distance—cities are brought into line, the unknown rendered both knowable and comparable (chapter 2). Perhaps the most well-known example of this is benchmarking. This process reduces urban complexity to a series of numbers, bringing into comparative coexistence territories from around the world. Cities are then ranked alongside one another (McCann 2004). As a result, the act of comparison becomes a particularly political act (was it ever really anything other?). Robinson (2006) has made this point in her work. She has argued that the work done by comparing different cities around the world on the basis that some cities are "global" and others are not can be profoundly disabling for those named "nonglobal."As a consequence, "'local' policy development now occurs in a self-consciously comparative . . . context" (Peck 2003, 229). Key Performance Indicators, which are the most common and globally well-known benchmarking technology, are no longer just the talk of corporate managers. Those running cities have increasingly found themselves governed at a distance by the disciplining consequences of various Key Performance Indicators. They make comparing and ranking easier. From crime rates to educational achievements, employment rates to environmental emissions, cities are now required to produce a growing amount of information about their performances across a range of different policy areas. The example of Business Improvement Districts is no exception. Key Performance Indicators include car park usage, footfall figures, vacancy property rates, and retail sales (Hogg et al. 2006). In the next section, I turn to the context of the Business Improvement District model.

Business Improvement Districts: Context

Rising to prominence in the early 1990s, the Business Improvement District concept is about both a way of governing space and an approach to its planning and regulation. A BID is a public–private partnership in which property and business owners in a defined geographical area vote to make a collective contribution to the maintenance, development, marketing, and promotion of their commercial district. So a Business Improvement District delivers advertising, cleaning, marketing, and security services across its geographical jurisdiction. Businesses vote to tax themselves to take management control over "their" area. Business Improvement Districts reflect how "property owners . . . , developers and builders, the local state, and those who hold the mortgage and public debt have much to gain from forging a local alliance to protect their interests and to ward off the threat of localized devaluation" (Harvey 1989, 149). BID proponents critique the past role of government in the business of governing the downtown. Instead, Business Improvement Districts are portrayed as a "more focused and flexible form of governance than large municipal bureaucracies" (Levy 2001, 129). Channeling "private sector agency towards the solution of public problems" (MacDonald 1996, 42), they are represented as "an alternative to traditional municipal planning and development" (Mitchell 2001, 116). Mallett (1994, 284) goes as far as to claim that Business Improvement Districts are "a response to the failure of local government to adequately maintain and manage spaces of the post-industrial city." The BID philosophy is that "the supervision of public space deters criminal activity and the physical design of public space affects criminal activity" (Hoyt 2004, 369). It draws on the work of Jacobs (1961), Newman (1972), and Wilson and Kelling (1982), which argued that the design of urban space could change the way people behave. As Business Improvement Districts establish the physical layout of benches, streetlighting, and shop facades, so they shape the ways in which an area is experienced. As such, the BID model draws on, and reinforces, contemporary neoliberal thinking on both the need to attend to and emphasize the "business climate" and the "quality of life," resonating with, if not predating, the much maligned "creative classes" work of Richard Florida (2002; see chapter 3).

The global diffusion of the Business Improvement District model since the mid-1990s has involved a number of agents of transference. Variously situated policy actors have been party to the international spread of this model. As it has been moved around the globe from one place to another, the model has been subject to a number of changes in its institutional DNA. As it has been territorialized—embedded in particular sociospatial relations—certain elements of the model have been emphasized, while others have been downplayed. A number of places along the way have shaped the model's form on its introduction into England and, most notably, New York (Ward 2006; McCann and Ward 2010). However, we begin by retracing its initial emergence.

Business Improvement Districts on the Move

The first BID was established in Toronto in 1970, and the model spread rapidly, encouraged by Canadian state funding incentives. After moving across Canada, it entered the United States, where the initial BID was set up in New Orleans in 1975. Quite how this occurred is not clear (Hoyt 2006). During the 1980s and 1990s, the number of U.S. Business Improvement Districts grew slowly but surely. Latest data suggest there are over five hundred across the country, with the majority in just three of the fifty states: California, New York, and Wisconsin (Mitchell 2001). However, this figure is now a little out of date and anecdotally it is likely there are more than a thousand U.S. Business Improvement Districts.

By the end of the 1990s, the Business Improvement District model in the United States had moved to center stage in local governments' efforts to revitalize their downtowns. Examples include activities in cities such as in Appleton, Wisconsin (Ward 2010), Atlanta, Georgia (Morçöl and Zimmermann 2008), Los Angeles, California (Meek and Hubler 2008), Milwaukee, Wisconsin (Ward 2007b), and San Diego, California (Staeheli and Mitchell 2007; Stokes 2008). Business Improvement Districts have then been established in a variety of U.S. localities. However, there is one place, above all others, whose experiences have been the basis for the internationalization of the BID model: Manhattan, New York.

Business Improvement Districts were first established in New York in the mid-1980s. However, in the early 1990s, as part of the Manhattan Institute's wider "new urban paradigm" (Magnet 2000) and Mayor Giuliani's "quality of life" campaign, the model rose to local and, eventually, international prominence (Ward 2006). As elements of these were internationalized, most notably "zero-tolerance" public safety policies (Wacquant 1999), the BID model, à la New York, began to attract the attention of various agents of transference, within the city and beyond. The Manhattan Institute's *City Journal* provided the intellectual rationale for this model of restructuring, flanked by a small number of books on high-profile issues such as health care reform and welfare restructuring. Regular breakfasts, conferences, and seminars brought people together. Columnists in the *City Journal* regularly critiqued city government, offering right-of-center alternatives and pushing a promarket, neoliberal agenda (Peck 2006). The BID model appeared to resonate with the Manhattan Institute's and the city government's wider political project. As one of its writers noted,

> [Business Improvement Districts] have returned to an earlier set of values regarding public space. They understand that simple things—such as keeping sidewalks clean and safe—matter enormously to the urban quality of life. A city that has lost the will to control allegedly "minor" offenses such as trash and graffiti invites further disorder. (MacDonald 1998, 2)

Other local journalists were equally gushing:

> Any discussion of BIDs must begin with a restatement of what they've accomplished. They have cleaned up, almost alone, key areas of Manhattan business districts. They have contributed to the drop in crime. Outside of Manhattan, they have undertaken crucial commercial refurbishing and marketing efforts. They have been able to do all this because they have *operated outside city government and with a fair amount of entrepreneurship*. (Lentz 1998, 4, emphasis added)

In light of the favorable reporting of its achievements, the New York BID model began to be looked at by all manner of "policy tourists" (Ward 2007a). The city's economic success was in part attributable to the role of Business Improvement Districts. Figures such as Daniel Biedermann were given an elevated status. He was the president of the Grand Central Partnership Social Services Corporation (GCSSC) that oversaw three of the largest Business Improvement Districts—Bryant Park, Grand Central Partnership, and Thirty-fourth Street. He became dubbed locally as the Mayor of Midtown. And, as over the past decade, the model has emerged in cities in Australia, Japan, Serbia, South Africa, and the United Kingdom (in design if not always in name), so senior executives, such as Daniel Biederman and Paul Levy, a past president of the downtown Philadelphia Business Improvement District, have become international BID "gurus." They have traveled around the world, writing guidance documents based on their experiences and giving practitioner and policymaker presentations. More generally, the diffusion of the Business Improvement District model has taken place through a number of channels, some relatively formal, others less so. The International Downtown Association (IDA)—physically located in Washington, D.C., and the center of a network of national downtown trade associations and convener of an annual conference—has been important to the BID model's internationalization. As it puts it,

> Founded in 1954, the International Downtown Association has more than 650 member organizations worldwide including: North America, Europe, Asia and Africa. Through our network of committed individuals, rich body of knowledge and unique capacity to nurture community-building partnerships, IDA is a guiding force in creating healthy and dynamic centers that anchor the well being of towns, cities and regions of the world. (IDA 2006a)

In its view, the BID model is one of the most successful ways to improve the conditions of downtowns the world over. According to the current IDA president, David Feehan, "the IDA is proud of the role it has played in the resurgence of downtowns in the US and Canada. Now, through partnerships in Europe, the Caribbean, Australia and Africa, IDA is expanding its resources and knowledge base even more" (IDA 2006b). Its partners include the Association of Town Centre Management (ATCM) in the UK, Business Improvement Areas of British Columbia (BIABC) in Canada, Caribbean Tourism Organization (CTO) in the West Indies, and Central Johannesburg Partnership (CJP) in South Africa. Neil Fraser, the executive director of the CJP, describes the role of the IDA as "a true leader in bringing together city practitioners and specialists from North America and around the world. They provide essential support and assistance in all aspects of private urban management" (IDA 2006c).

Through regular conferences, institutes, seminars, and workshops organized by the IDA, downtown practitioners have fed into and reinforced the general emphasis on creative and livable cities (Florida 2002). Together with national partners and others with a stake in the expansion of the BID model, such as private consultancies, think tanks, and government departments, activities of the IDA have convinced urban authorities of the virtues of the BID model. In 1995, the CJP and IDA organized a "study tour" to the UK and the United States for Johannesburg's public and private sector officials. The purpose was "to visit . . . sites and learn from international experiences in order to set up practices and legislation for a CID [City Improvement District] in Johannesburg" (Peyroux 2008, 4). This type of activity is symptomatic of the means through which education occurs—a mixture of off-the-peg learning blended with real-life, in-your-face evidence (Wolman 1992; Peck 2003; McCann 2008).

Less formally, but no less important in the model's internationalization, have been figures involved in the BID model in some of the largest East Coast U.S. cities (Ward 2006; Cook 2008). As already mentioned, Biederman and Levy in particular, have worked hard to promote the BID model around the world. According to Peyroux (2008, 4), "the North American BIDS [Business Improvement Districts] were a strong reference for the Johannesburg CIDs." The two of them have presented in many countries, extolling its virtues, drawing on their own highly situated and quite specific experiences to "market" the model and its benefits. Various exchange-making and information-sharing events have been organized in cities, including Canberra, Dublin, Johannesburg, London, and Newcastle, Australia. At these, an ever-wider audience of different types of practitioners and policymakers has been educated in the way of Business Improvement Districts. Not only development officers and planning officials, as might be expected, attend and participate at these events. Because of the financial and legal consequences of BID formation, accountants and lawyers are also selected into the web of mobilization. They learn about the financial and legal consequences of the BID model (Ward 2006).

When organizing local events, agents of transference have tailored general lessons to the specific concerns of host countries or cities. The trick to the ongoing global diffusion of this model of downtown governance has been, of course, to ensure that assembled audiences are convinced both of the virtues of the BID model in general and of its capacity to attend to whatever issues a particular local representative may be facing. In England, the particular case to which this chapter now turns, this has meant marketing the BID model in the context of an already extant town management system.

Business Improvement Districts in Place

In addition to the supply side, a demand side is needed for policy transfer to occur, although these need to be understood as mutually constituted and reinforcing. Locally dependent or embedded agents of transference play an important role in translating a general model into something that makes sense to those with territorial remits. They act as intermediaries, directly, in some cases, through acts of translation such as conference talks or indirectly, through assembling, preparing, and stabilizing knowledge. An example would be the Knowledge Bank established by http://www.ukbids.org. This is a virtual resource for those involved in the process of introducing the BID model into England. In it are assembled case studies from around the world, territorialized knowledge rendered both mobile and fixed simultaneously. Best-practice examples of different aspects of the BID model, from suggestions about how to establish a BID through to ways of marketing and promoting downtowns, are assembled in the bank and can be withdrawn by interested parties.

In the case of the UK, the introduction of the BID model was first mooted in the early 1990s. As the latest in a long line of post–Second World War exchanges of urban policies between the two countries (Barnekov et al. 1989; Wolman 1992; Peck and Theodore 2001; Wacquant 2001; Jonas and Ward 2002), a report commissioned by the Corporation of London considered the lessons the city might learn from the BID model in New York (Travers and Weimar 1996). Although this report argued for the model's introduction into London, it reflected a clearer and more general push among town center practitioners to do something about the management of downtowns. This debate was led by the Association of Town Centre Management. However, it was not until after the election of the national Labour government in 1997 that a series of "urban" policy documents of different sorts were issued, most noticeably Lord Rogers's *Towards an Urban Renaissance* (DETR 1999, 2005) and the government's *White Paper. Our Towns and Cities: The Future* (DETR 2000). These focused political and practitioner attention on the role cities should be encouraged to play in driving national economic growth. All were informed by examples of policy tourism. Government ministers, such as John Prescott, and senior officials were regular visitors to New York and Philadelphia. They were keen to see the BID model in action. A series of other green and white papers were issued at the end of the 1990s and the beginning of the 2000s to create the financial and legal conditions in English cities for the creation of Business Improvement Districts.

During this period, there was a sustained creation of favorable "importing" conditions. Various agents of transference operating at and across a range of "spaces of engagement" (Cox 1998), such as national think tanks, regional development agencies, and local authorities, lobbied on behalf of the establishment of Business Improvement Districts. A series of documents were produced and circulated. Reports appeared in trade association magazines, such as *Regeneration and Renewal* and *Town and Country Planning,* and on trade Web sites, such as http://www .publicfinance.co.uk. As we have already seen, a made-to-order Web site— http://www.ukbids.org—was established to oversee the introduction of Business Improvement Districts into England (with Scotland and Wales expected to follow shortly afterward). In its words,

> UKBIDs is committed to supporting robust and successful Business Improvement Districts (BIDs) in the United Kingdom. UKBID

incorporates the National BIDs Advisory Service and is delivered by the Association of Town Centre Management, who led the government-supported National BIDs Pilot that introduced BIDs to England and Wales. Today we work actively with new and established BIDs across the country, and with strategic organizations such as the Regional Development Agencies. We lead the national BID Network Exchange and are delivering the country's first BIDs Academy, as well as undertaking research, training events and seminars. (http://www.ukbids.org)

Jacqueline Reilly was appointed as the project director of the National Business Improvement Pilot Project (and subsequently to run its successor, the UKBIDs Advisory Service). She championed the BID model in England, acting as both expert and advocate (Rich 2004). The role of ukbids .org as a clearinghouse for pro-BIDs expertise, linking Britain to the global discussion, is clear from the Web site:

> Through our relationship with ATCM (Association of Town Centre Management), and the unique reciprocal membership scheme with the International Downtown Association (IDA) based in Washington DC, our BID network is the largest BIDs network in the world and our Knowledge Bank an unrivalled resource for information on both BIDs and partnership development. Building on our own experience from the National BIDs Pilot, the Knowledge Bank is growing all the time, as members exchange expertise in the BID Network Exchange and other partnership events across the country. (http://www.ukbids.org)

The creation of English and Welsh Business Improvement Districts was finally announced in 2001, and the final piece of the legal framework was agreed in 2004. Scotland subsequently followed suit. Despite its Canadian origins, it was the United States that was named publicly as the geographical reference point for the model, with Business Improvement Districts hailed as "New York–style schemes" (ODPM 2003, 1): "I can tell you today that we have decided to introduce legislation to create Business Improvement Districts. These will be similar to the successful US examples" (DETR 2001, 1). "This approach [to the BID program] building on the very successful business model in the USA, will allow businesses to see precisely what they are getting for their money and will help to harness local business leadership" (DETR 2001, 2). Of course, the BID model was not introduced into an institutional vacuum in England. Around the country, many cities and towns had already in place some sort of governing partnership. Many hundreds had town center management partnerships recognized by the Association of Town Centre Management (Reeve 2004, 2008). These shared many characteristics with the BID model, bar an important one: businesses made their financial contribution voluntarily.

This meant that contributions could fluctuate year after year, undermining the capacity of the partnerships to plan in the medium term (Cook 2008). Indeed, the International Downtown Association's first annual conference took place in Coventry, England, in 1997. So the experiences of some of England's cities were already present in the geographical imagination of international practitioners. In addition, the public finance system in England remains highly centralized. There are few examples of city government raising revenue through taxes. And, as Peck and Theodore (2001, 430) remind us, "inherited institutional structures, established political traditions, and extant policy conventions and discourses all operate to ensure a degree of continuity in the policy development process." In the case of the BID model, this matters nationally and locally. The centralized system of central-local government relations affects the way something like the BID model would be introduced. In different localities, it is important that those involved in mediating and translating the BID model are aware of its particular issues. Put simply, while there is much that unites Bolton, Brighton, and Coventry, there is also much that distinguishes them.

Unsurprisingly, then, the English BID model as established, and as it has been enacted from locality to locality, is quite unique. In particular, it differs in three quite fundamental ways in design from the U.S.-derived model that has done the rounds internationally. First, this was a state-sponsored introduction of the BID model, an example of the role the nation-state can play in co-coordinating, managing, and regulating new modes of governance. In this context, "the national level assumes responsibility for coordinating activities of local partnerships and model delivery systems and for establishing the rules of the game, while the local level—the scale of innovation and implementation—plays a decisive role in translating national policies and local lessons into practice" (Peck and Theodore 2001, 432–33). English cities and towns competed for a place in the National Business Improvement District Pilot Project. More than one hundred applied and twenty-three were successful. These were pilot Business Improvement Districts that ran for a couple of years while cities prepared themselves for a vote. Since the ending of the pilot scheme, any city or town in England has been able to hold a vote. This takes us to the second peculiarity of the English BID model. In the United States, property owners vote. In the UK, all nondomestic ratepayers, that is, those who rent properties, vote in the BID referendum. This was the outcome of a long debate among vested interests—local and national government, retail trade associations, property owners, and so on (Cook 2008). Despite evidence of involvement by property owners in the activities of Business Improvement Districts, this does not stretch to getting a vote in their establishment. Third, a successful vote must past two tests: more than 50 percent of the votes cast must favor the BID, and the positive vote must represent more than 50 percent of the rateable value of the votes cast.[1] So there are particular politics around the local dependency of businesses (Cox and Mair 1988). Territory still matters. In some instances, the first criterion has been met, but the second has not, as typically smaller, local independent businesses may have voted "yes," while multisite chains, which are typically larger and hence have a higher rateable value, have voted "no." By the end of 2008, England had seventy Business Improvement Districts (Table 4.1). Five things are worth noting. First, there is no clear underlying geographical pattern to the establishment of English Business Improvement Districts. They have been created in cities and towns of all sizes and all parts of the country. Second, voter turnout has been consistently low. There have been exceptions, of course, but in general, businesses have not turned out en masse. In most cases, less than half the eligible businesses voted. Fourth, seventeen votes were unsuccessful the first time around. Two have subsequently been established, although in the cases of Maidstone, Runnymede, and Southport, there have been two unsuccessful votes. Fifth, there have been successful second-term ballots in Bristol, Coventry, Liverpool, and London (2). In these cases, the BID was initially constituted for three years, and the management committee subsequently sought a second term of office.

So there have been a variety of issues around the introduction of the BID model into English localities. This highlights the complicated ways in which the BID model has been both moved around the world and embedded in existing territorially constituted social relations. It has been moved from one city to another through a myriad of formal and informal networks, through the procedural and technocratic transfer of policy on the one hand, and the presentational performances of high-profile individuals on the other. Simultaneously and necessarily, the BID model has been embedded or "fixed" temporarily in national and local contexts through the activities of a set of territorially entangled agents of transference. It is a policy model with necessary relational and territorial elements. These not surprisingly have produced a hybrid version of the New York "model," which itself is a variation on the original Canadian example.

Conclusion

In a number of English localities, BID votes are imminent. Royston will not be the last. According to Hoyt (2006, 433), "policy entrepreneurs in countries around the globe continue to advocate the transfer of BID

Table 4.1. England's Business Improvement Districts (December 2008)

BID	Ballot date	Yes vote by number of businesses (%)	Yes vote by rateable value of businesses (%)	Turnout (%)
Royston First	12/02/08	61	62	39
Preston	11/28/08	73	83	25
Newcastle	11/24/08	67	59	52
Hinckley	11/18/08	64	70	39
Bristol Broadmead (2nd-term ballot)	10/31/08	55	55	53
Paddington (2nd-term ballot)	10/30/08	83	90	46
Boston	10/22/08	73	83	28
Liverpool (2nd-term ballot)	10/17/08	64	68	42
Leamington	03/31/08	61	63	41
Bathgate	03/14/08	93	82	45
Daventry First	03/13/08	74	80	27
Coventry City Centre (2nd-term ballot)	02/29/08	83	85	36
Dorchester BID Co.	02/29/08	81	84	56
New West End Co. (London) (2nd-term ballot)	12/21/07	63	73	43
Astmoor Industrial Estate	12/06/07	72	77	65
Halebank Industrial Estate	12/06/07	72	70	50
Derby Cathedral Quarter	11/28/07	85	74	43
Longhill and Sandgate (Hartlepool)	11/12/07	80	94	29
Nottingham Leisure	10/26/07	75	75	33
Kings Heath	08/28/07	74	53	27
Blackburn EDZ Industrial Estate	08/02/07	89	89	40
Taunton	07/31/07	72	67	42
Winchester	07/26/07	54	62	45
Worthing Town Centre	07/05/07	57	53	31

(continued on next page)

Table 4.1. England's Business Improvement Districts (continued)

BID	Ballot date	Yes vote by number of businesses (%)	Yes vote by rateable value of businesses (%)	Turnout (%)
Sleaford	07/05/07	69	75	40
E11	06/22/07	95	91	42
Argall	05/23/07	86	93	TBC
Segensworth Estates (Fareham)	07/15/07	73	TBC	30
Cannock Chase	03/30/07	62	68	44
Erdington	03/29/07	74	55	31
Croydon	02/28/07	63	70	44
London Riverside	02/26/07	82	68	30
Heart of London Business Alliance (2nd-term ballot)	02/26/07	86	89	62
Angel Town Centre	02/23/07	77	83	51
Coventry City Wide	02/22/07	54	59	33
Cater Business Park	02/05/07	90	80	56
InSwindon	02/01/07	69	54	41
Oldham	12/06/06	76	56	45
Southern Cross	12/04/06	94	99	72
Retail Birmingham	11/09/06	69	62	49
Altham (2nd ballot)	11/08/06	61	70	70
Hull	10/18/06	81	76	45
Cowpen Industrial Association	10/05/06	88	87	32
Ipswich	07/24/06	66	70	49
Brighton	05/26/06	64	70	46
Swansea	05/04/06	74	65	45
West Bromwich	04/07/06	79	85	48
Hammersmith	03/29/06	57	70	48
Great Yarmouth	03/28/06	82	88	44
Ealing	03/28/06	65	64	51
Hainault Business Park	03/20/06	85	93	52
Camden Town Unlimited	03/01/06	83	84	50
Waterloo Quarter Business Alliance	03/01/06	74	92	50
Bolton Industrial Estates	12/01/05	72	84	46

(continued on next page)

Table 4.1. England's Business Improvement Districts (continued)

BID	Ballot date	Yes vote by number of businesses (%)	Yes vote by rateable value of businesses (%)	Turnout (%)
Reading	11/19/05	89	71	50
London Bridge	11/17/05	71	78	50
Liverpool City Central (2nd ballot)	10/20/05	62	51	56
Rugby	09/30/05	66	74	50
Keswick	09/22/05	55	74	50
Blackpool Town Centre	08/23/05	89	74	40
Bristol Broadmead	06/30/05	60	56	59
Birmingham Broad Street	05/26/05	92	97	65
Lincoln	04/18/05	79	83	44
Bedford	03/30/05	77	81	40
New West End Co. (London)	03/16/05	61	69	53
Plymouth	03/01/05	77	66	58
Paddington (London)	03/01/05	87	88	51
Coventry City Centre	02/24/05	78	75	38
Holborn Partnership (London)	02/11/05	82	77	50
Better Bankside (London)	01/24/05	75	67	48
Heart of London Business Alliance	12/31/04	71	73	62
Kingston First	11/16/04	66	66	TBC

Source: (UKBIDs, n.d.)

policy." A number of examples exist of cities with quite diverse backgrounds adopting the BID model, in spirit if not in name. It has become the downtown revitalization method of choice. The extent to which the model "works" is moot. There is no shortage of consultancy reports that claim to demonstrate their successes, period. A whole series of data are marshaled to make the point, of course increasing the likelihood that the model be will kept moving. Nevertheless, while there is some evidence that they increase the value of the capital invested in the built environment, they do this at a cost regarding issues of citizenship, democracy, and the future of public space in the city (Staeheli and Mitchell 2007; Ward 2007a).

This chapter has revealed how the Business Improvement District model has been moved around the world and has outlined its geographical and ideological origins. There is a difference. The model began its life in Canada. Its ideological origins have been manufactured in the United States, as U.S. cities, most notably New York, have become the reference points for cities in other countries. The chapter then turned to documenting those actors and institutions involved in the transnationalization of the BID model. A range of agents of transference were revealed. Some with little reach, overseeing its introduction in a specific city, such as the example of Royston with which I began this chapter. Others with a far longer reach, able to influence policy reform at a distance, such as the model's U.S. gurus whose trans-Atlantic visits were important in facilitating the introduction of the BID model into the UK. The chapter then moved on to examine the ways in which the BID model was introduced into England, reaffirming the extent to which urban development remains shaped by path-dependent forces. It reveals the ways in which a process of translation is performed, both by those coming in from outside and by actors resident in each of the contexts. In these moments—whether they are literally "performed" at conferences or workshops, or occur through circulated written publications—supply and demand come so close as to be almost indistinguishable.

While these empirical details are important, this chapter concludes by describing three conceptual issues that speak to the wider objectives of this edited collection. First, this chapter has argued for an appreciation of how cities are assembled by the situated practices and imaginations of actors who are continually attracting, managing, promoting, and resisting global flows of policies and models (chapter 5). The bringing together of policies and models from around the world of today constitutes the path dependency trajectories of tomorrow. Second, it has advanced a framework that includes a broad understanding of those involved in the mobility of policies, takes seriously the transfer of interurban, transnational models, understanding "transfer" as a sociospatial process in which models are subject to change as they are moved (chapters 2 and 5). The New York case is particularly revealing. The BID model that left New York for England was not the BID model that was introduced into New York at the end of the 1980s. By the mid-1990s, when it caught the attention of those with a stake in the sustainability of England's city and town centers, the model had been put to work by those in New York pursuing what Smith (1996) terms "revanchist urbanism." Third, the approach developed in this chapter has, at its core, a sensitivity to both structure and agency. In the case of the BID model, certain individuals did make a difference. This was not done under terms of their own making, however. Rather, there is a set of macro supply and demand contexts in which some are structurally advantaged (chapter 3). Some, more than others, are likely to have their ideas and policies made mobile. Of course, there is an interaction of a range of differently scaled forces in and through which these agents mobilize, broker, translate, and introduce ideas in such a way as to make the territorially embedding of policies and models not just possible but probable.

Note

1. In the United States, there is no federal or statewide voting system for the creation of a BID. It differs within states, and even within cities (Ward 2007b).

References

Allen, J., and A. Cochrane. 2007. "Beyond the Territorial Fix: Regional Assemblages, Politics, and Power." *Regional Studies* 41:1161–75.

Allen, J., D. Massey, and A. Cochrane. 1998. *Re-Thinking the Region*. London: Routledge.

Amin, A. 2004. "Regions Unbound: Towards a New Politics of Place." *Geografiska Annaler* 86B:33–44.

Barnekov, T., R. Boyle, and D. Rich. 1989. *Privatism and Urban Policy in Britain and the United States*. Oxford: Oxford University Press.

Blaut, J. M. 1993. *1492: The Debate on Colonialism, Eurocentrism, and History*. Trenton, N.J.: Africa World Press.

Brenner, N. 2004. *New State Spaces: Urban Governance and the Rescaling of Statehood*. Oxford: Oxford University Press.

Campbell, J. L. 2001. "Institutional Analysis and the Role of Ideas in Political Economy." In *The Rise of Neoliberalism and Institutional Analysis*, edited by

J. Campbell and O. K. Pederson, 159–89. Princeton, N.J.: Princeton University Press.

Cook, I. R. 2008. "Mobilising Urban Policies: The Policy Transfer of U.S. Business Improvement Districts to England and Wales." *Urban Studies* 444:773–95.

Cox, K. R. 1995. "Globalisation, Competition, and the Politics of Local Economic Development." *Urban Studies* 32:213–24.

———. 1998. Spaces of Dependence, Spaces of Engagement, and the Politics of Scale, or: Looking for Local Politics." *Political Geography* 17:1–24.

Cox, K. R., and A. Mair. 1988. "Locality and Community in the Politics of Local Economic Development." *Annals of the Association of American Geographers* 78:307–25.

Department of the Environment, Transport and the Regions (DETR). 1999. *Towards an Urban Renaissance: Final Report of the Urban Task Force.* London: DETR.

———. 2000. *Our Towns and Cities. The Future: Delivering an Urban Renaissance.* London: The Stationary Office.

———. 2001. "Blair Unveils BIDs: A Scheme to Improve Local Quality of Life." News release, 2001/0234.

———. 2005. *Towards a Stronger Urban Renaissance.* http://www.urbantask force.org/ UTF_final_report.pdf.

Florida, R. 2002. *The Rise of the Creative Class: And How It's Transforming Work, Leisure, Community, and Everyday Life.* New York: Basic Books.

Hall, P., ed. 1989. *The Political Power of Economic Ideas.* Princeton, N.J.: Princeton University Press.

Harvey, D. 1989. "From Managerialism to Entrepreneurialism: The Transformation in Urban Governance in Late Capitalism." *Geografiska Annaler Series B* 71:3–17.

Healey, P., and R. Upton, eds. 2010. *Crossing Borders: International Exchange and Planning Practices.* Routledge: London.

Hogg, S., D. Medway, and G. Warnaby. 2006. "Town Centre Management Schemes in the UK: Marketing and Performance Indicators." *International Journal of Non-Profit and Voluntary Sector Marketing* 9:309–19.

Hoyt, L. 2004. "Collecting Private Funds for Safer Public Spaces: An Empirical Examination of the Business Improvement Districts Concept." *Environment and Planning B: Planning and Design* 31:367–80.

———. 2006. "Importing Ideas: The Transnational Transfer of Urban Revitalization Policy." *International Journal of Public Administration* 29:221–43.

Industrial Downtown Association (IDA). 2006a. "About IDA." http://www.ida -downtown.org/eweb/DynamicPage.aspx?webcode=aboutIDA.

———. 2006b. "From the President." http://www.ida-downtown.org/eweb/Dynamic Page.aspx?Site=IDA&WebKey=4e438326-d510-426f-a001-db13bf75c1c0.

———. 2006c. "Membership." http://www.ida-downtown.org/eweb//DynamicPage .aspx?WebKey=CF943A0F-88CE-4E9F-8B42-0A96B8989894.

Jacobs, J. 1961. *The Death and Life of Great American Cities.* New York: Vintage Books.

Jonas, A. E. G., and K. Ward. 2002. "A World of Regionalisms? Towards a U.S.–UK Urban and Regional Policy Framework Comparison." *Journal of Urban Affairs* 24:377–401.

Jonas, A. E. G., and D. Wilson, eds. 1999. *The Urban Growth Machine: Critical Perspectives, Two Decades Later.* Albany: State University of New York Press.

Larner, W. 2003. "Guest Editorial: Neoliberalism?" *Environment and Planning D: Society and Space* 21:508–12.

Larner, W., and R. Le Heron. 2002. "From Economic Globalization to Globalizing Economic Processes: Towards Post-Structural Political Economies." *Geoforum* 33:415–19.

Lauria, M., ed. 1997. *Reconstructing Urban Regime Theory.* Thousand Oaks, Calif.: Sage.

Lentz, E. 1998. "Business Improvements Districts Do the Business." *New York Times*, March 15, 4.

Levy, P. R. 2001. "Paying for Public Life." *Economic Development Quarterly* 15:124–31.

Logan, J., and H. Molotch. 1987. *Urban Fortunes: The Political Economy of Place.* Berkeley: University of California Press.

MacDonald, H. 1996. "Why Business Improvements Districts Work." *Civic Bulletin* 4. http://www.manhattan-institute.org/html/cb_4.htm.

———. 1998. "Why Business Improvement Districts Work." *Civic Bulletin* 4:1–4.

MacLeod, G., and M. R. Jones. 2007. "Territorial, Scalar, Networked, Connected: In What Sense a 'Regional World'?" *Regional Studies* 41:1177–91.

Magnet, M., ed. 2000. *The Millennial City: A New Urban Paradigm for 21st Century America.* New York: Ivan R. Dee.

Mallett, W. J. 1994. "Managing the Post-Industrial City: Business Improvement Districts in the United States." *Area* 26:276–87.

Massey, D. 1993. "Power-Geometry and a Progressive Sense of Place." In *Mapping the Futures: Local Cultures, Global Change*, edited by J. Bird, B. Curtis, T. Putman, G. Robertson, and L. Tickner, 59–69. New York: Routledge.

———. 1999. "Imagining Globalisation: Power-Geometries of Space-Time." In *Global Futures: Migration, Environment, and Globalisation*, edited by A. Brah, M. Hickman, and M. MacanGhaill, 27–44. Basingstoke, UK: St. Martin's Press.

———. 2005. *For Space.* London: Sage.

———. 2007. *World City.* Cambridge: Polity Press.

McCann, E. J. 2004. "'Best Places': Interurban Competition, Quality of Life, and Popular Media Discourse." *Urban Studies* 41:1909–29.

———. 2008. "Expertise, Truth, and Urban Policy Mobilities: Global Circuits of Knowledge in the Development of Vancouver, Canada's "Four Pillar" Drug Strategy." *Environment and Planning A* 40:885–904.

———. 2011. "Urban Policy Mobilities and Global Relational Geographies: Toward a Research Agenda." *Annals of the Association of American Geographers* 101:107–30.

McCann, E. J., and K. Ward. 2010. "Relationality/Territoriality: Toward a Conceptualization of Cities in the World." *Geoforum* 41:175–84.

McFarlane, C. 2011. *Learning the City: Knowledge and Translocal Assemblage.* Oxford: Wiley-Blackwell.

Meek, J. W., and P. Hubler. 2008. "Business Improvement Districts in Los Angeles Metropolitan Area: Implication for Local Governance." In *Business Improvement Districts: Research, Theories, and Controversies,* edited by G. Morçöl, L. Hoyt, J. W. Meek, and U. Zimmermann, 197–220. Boca Raton, Fla.: Auerbach Publications.

Mitchell, J. 2001. "Business Improvement Districts and the Management of Innovation." *American Review of Public Administration* 31:201–17.

Morçöl, G., and U. Zimmermann. 2008. "Community Improvement Districts in Metropolitan Atlanta." In *Business Improvement Districts: Research, Theories, and Controversies,* edited by G. Morçöl, L. Hoyt, J. W. Meek, and U. Zimmermann, 349–72. Boca Raton, Fla.: Auerbach Publications.

Morgan, K. 2007. "The Polycentric State: New Spaces of Empowerment and Engagement?" *Regional Studies* 41:1237–51.

Newman, O. 1972. *Defensible Space: Crime Prevention through Urban Design.* New York: Macmillan.

Office of the Deputy Prime Minister (ODPM). 2003. "Businesses 'BID' to Help Communities Thrive." News release 2003/0005.

Peck, J. 1995. "Moving and Shaking: Business Elites, State Localism, and Urban Privatism." *Progress in Human Geography* 19:16–46.

———. 2003. "Geography and Public Policy: Mapping the Penal State." *Progress in Human Geography* 27:222–32.

———. 2006. "Liberating the City: Between New York and New Orleans." *Urban Geography* 27:681–723.

Peck, J., and N. Theodore. 2001. "Exporting Workfare/Importing Welfare-to-Work: Exploring the Politics of Third Way Policy Transfer." *Political Geography* 20:427–60.

Peyroux, E. 2008. "City Improvement Districts in Johannesburg: An Examination of the Local Variations of the BID Model." In *Business Improvement Districts, Ein neues Governance-Modell aus Perspektive von Praxis und Stadtforschung, Geographische Handelsforschung, 14,* edited by R. Pütz, 139–62. Würzburg, Germany: Institut für Geographie und Geologie.

Prince, R. 2010. "Policy Transfer as Policy Assemblage: Making Policy for the Creative Industries." *Environment and Planning* A 42: 169–86.

Reeve, A. 2004. "Town Centre Management: Developing a Research Agenda in an Emerging Field." *Urban Design International* 9:133–50.

———. 2008. "British Town Center Management: Setting the Stage for the BID Model in Europe." In *Business Improvement Districts: Research, Theories and Controversies*, edited by G. Morçöl, L. Hoyt, J. W. Meek, and U. Zimmermann, 350–423. Boca Raton, Fla.: Auerbach Publications.

Rich, A. 2004. *Think Tanks, Public Policy, and the Politics of Expertise.* Cambridge: Cambridge University Press.

Robinson, J. 2006. *Ordinary Cities: Between Modernity and Development.* London: Routledge.

Robinson, J. 2008. "Developing Ordinary Cities: City Visioning Processes in Durban and Johannesburg." *Environment and Planning A* 40: 74–87.

Royston First. 2008. "Frequently Asked Questions about the Royston Bid." http://www.woodenhouse.com/roystonfirst/Resources/FAQs.pdf.

Saunier, P.-Y. 2002. "Taking Up the Bet on Connections: A Municipal Contribution." *Contemporary European History* 11:507–27.

Sayer, A. 2004. "Review Article: Seeking the Geographies of Power." *Economy and Society* 33:255–70.

Smith, N. 1996. *The New Urban Frontier: Gentrification and the Revanchist City.* London: Routledge.

Staeheli, L., and D. Mitchell. 2007. *The People's Property: Power, Politics, and the Public.* London: Routledge.

Stokes, R. J. 2008. "Business Improvement Districts and Small Business Advocacy: The Case of San Diego." In *Business Improvement Districts: Research, Theories, and Controversies*, edited by G. Morçöl, L. Hoyt, J. W. Meek, and U. Zimmermann, 249–68. Boca Raton, Fla.: Auerbach Publications.

Stone, C. 1989. *Regime Politics: Governing Atlanta, 1946–1988.* Lawrence: University of Kansas Press.

Travers, T., and J. Weimar. 1996. *Business Improvement Districts: New York and London.* London: The Greater London Group, London School of Economics and Political Science.

UKBIDs, the National BIDs Advisory Service. "Welcome to UK BIDs—the National BIDs Advisory Service." http://www.ukbids.org.

Wacquant, L. 1999. "How Penal Common Sense Comes to Europeans: Notes on the Transatlantic Diffusion of the Neoliberal Doxa." *European Societies* 1:319–52.

———. 2001. "The Penalisation of Poverty and the Rise of Neo-Liberalism." *European Journal on Criminal Policy and Research* 9:401–12.

Ward, K. 2000. "A Critique in Search of a Corpus: Re-Visiting Governance and Re-Interpreting Urban Politics." *Transactions of the Institute of the British Geographers* 25:169–85.

———. 2006. "'Policies in Motion,' Urban Management, and State Restructuring: The Trans-Local Expansion of Business Improvement Districts." *International Journal of Urban and Regional Research* 30:54–75.

————. 2007a. "Business Improvement Districts. Policy Origins: Mobile Policies and Urban Liveability." *Geography Compass* 2:657–72.

————. 2007b. "'Creating a Personality for Downtown': Business Improvement Districts in Milwaukee." *Urban Geography* 28:781–808.

————. 2010. "Entrepreneurial Urbanism and Business Improvement Districts in the State of Wisconsin: A Cosmopolitan Critique." *Annals of the Association of American Geographers* 100:1177–96.

Wilson, J., and G. Kelling. 1982. "'Broken Windows': The Police and Neighborhood Safety." *Atlantic Monthly*, March, 29–38.

Wolman, H. 1992. "Understanding Cross-National Policy Transfers: The Case of Britain and the U.S." *Governance: An International Journal of Policy and Administration* 5:27–45.

CHAPTER FIVE

Points of Reference

Knowledge of Elsewhere in the
Politics of Urban Drug Policy

Eugene McCann

On an early May evening in 2006, two meetings on drug policy and treatment took place within a mile and a half of each other, on either end of the Downtown Eastside neighborhood of Vancouver, British Columbia. The more publicized and well-attended meeting was part of a series of public dialogues entitled "Beyond Criminalization: Healthier Ways to Control Drugs." These were organized by local advocates and practitioners of harm reduction, an approach to drug use that considers it primarily a public health concern, rather than a crime. Harm reduction practitioners accept the reality of drug use, eschew the ideal of a drug-free society, and evaluate the effects of drug use along a continuum, ranging from uses that are less harmful to those that are extremely harmful to an individual or society. Abstinence, from this perspective, is an ideal outcome for some but is not the only acceptable behavior or goal nor the fundamental precondition for entry into treatment. It is an approach that is more open to the advocacy of drug users, that defines users as partners in their own care, and that seeks to reduce the stigma associated with addiction through the pragmatic and nonjudgmental "management of everyday affairs and actual practices . . . [the] validity [of which] is assessed by practical results" (Marlatt 1998, 56).

Harm reduction is most commonly associated with methadone prescription and needle exchange programs. Increasingly, it also involves the medical prescription of heroin as part of Heroin Assisted Treatment (HAT), which intends to remove users from harmful "street scenes" and thus begin stabilizing their lives and addressing their relationship with psychoactive substances (Marlatt 1998; Riley and O'Hare 2000). Furthermore, cities in eight countries[1] operate supervised facilities for the consumption of illicit drugs. This iteration of harm reduction provides a relatively safe space—as compared with streets, alleys, and so forth—in which to inject or otherwise consume drugs, including heroin and cocaine. Users bring their street-bought drugs, often with the encouragement of local police, to these facilities and staff provide sterile equipment and monitor users for signs of overdose. These supervised sites also act as "low-threshold" entryways into a range of medical treatments and social services (Marlatt 1998; Maté 2008).

After studying similar policies elsewhere, especially in Germany and Switzerland, Vancouver officially adopted a "four pillar" drug policy in 2001. It aligns harm reduction with enforcement, treatment, and prevention. In 2003, Insite, North America's only legal-supervised injection site, was opened in a Downtown Eastside storefront, and, from 2005 to 2008, the North American Opiate Medication Initiative (NAOMI) prescribed heroin to a select number of users in a nearby building. These initiatives would have been inconceivable a decade before and were the result of political pressure exerted by harm reduction advocates. This political element of harm reduction practice continues in cities across the globe because many jurisdictions still prohibit needle exchanges as well as supervised consumption sites and HAT programs. Activism is also broadening, as some call for the end of prohibition and the creation of regulated markets for currently illegal drugs, similar to markets for alcohol and tobacco (Haden 2004).[2] It was at this vanguard of harm reduction advocacy that Vancouver's Beyond Criminalization meetings were positioned. They were held in parallel with the annual conference of the International Harm Reduction Association (IHRA, pronounced "Ira"), featured many of IHRA's leading lights, including physicians, public health professionals, bureaucrats, activists, and drug users, and filled a 150-seat university conference facility in the city's downtown business district.

The other public meeting held that evening focused on a different approach to the problems of drug use. In the upstairs meeting room of a community center, a dozen people listened to a local community worker's presentation on San Patrignano, an abstinence-based drug treatment center located outside Rimini, Italy, to which he had traveled on a fact-finding visit. The presentation involved a detailed description of the San Patrignano "therapeutic community," which practices a strict abstinence-based treatment regime for drug users who spend long periods, usually years, at the rural facility, isolated from their former lives, learning various trades, and producing products that are sold, often at high prices, to subsidize their treatment (field notes, May 2, 2006). San Patrignano's operators argue that drug addiction can be cured completely in a relatively short period through intensive behavioral, rather than medical, treatments intended to make users behave responsibly. They report that 72 percent of those who have lived at the community for at least eighteen months have "fully recovered" (San Patrignano 2008a). Yet others have questioned these results and point to the strict admission requirements that tend to weed out those less likely to show success, particularly San Patrignano's prohibition on drug users with concurrent mental health diagnoses (San Patrignano 2008b). Critics also point to a history of harsh treatment of residents by their peers and by staff (Arnao, n.d.; McMartin 2006). While acknowledging these concerns as valid, the community worker advocated a San Patrignano–inspired model for British Columbia and distributed a business plan for such a community in a rural part of the province (interview, community worker, 2006). For him, long-term residential treatment—particularly with a skills training component—was a necessary correction to the harm reduction approach that had come to dominate Vancouver's drug policy.

My purpose in this chapter is to use the ongoing politics of drug policy in Vancouver to address a central question of this book: how do transfers of policy models across the globe change the character and conduct of urban politics once they are territorialized in a new location? I argue that "policy mobilities" (McCann 2008, 2011) frequently have long-term political consequences for cities, beyond the immediate negotiations over whether a policy from elsewhere should be emulated. New points of reference, beyond the immediate local and national political context, become embedded in local political debate through travel, representation, repetition, and contest, thus constituting urban politics as both territorial and global–relational. Therefore, it is important to analyze the ongoing resonance of exemplars from elsewhere in local politics. Another key question of this book is, can urban–global relations be usefully analyzed through detailed, theoretically attuned empirical case studies? I seek to show that they can, by drawing on an ongoing qualitative research project involving interviews with user-activists, other policy activists, senior politicians, public health workers, researchers, and police, direct observations at meetings, and archival research.

In the next section, I position my argument within contemporary literatures on the relationality of place and of politics, on urban policy mobilities, and on the governance of public health at and among various scales. Subsequently, I return to Vancouver to discuss in more detail the debate over evidence from elsewhere that occurred during the original campaign for harm reduction at the turn of the twenty-first century, focusing on the role of Frankfurt's and Zürich's harm reduction programs as points of reference in the struggle over policy change.[3] I also note how those debates continue to resonate in local politics today. The subsequent section discusses the period after the institution of the four pillars strategy (2001), the opening of Insite (2003), and the initiation of the NAOMI trial (2005), in which critics of harm reduction have increasingly invoked San Patrignano's therapeutic community model as another approach to addiction. The chapter ends by arguing that, as cities are assemblages of "parts of elsewhere" (Allen and Cochrane 2007), their politics include assemblages of disparate reference points that resonate long term and condition political discourse and policymaking practice to look globally for inspiration and legitimation (chapters 2 and 4).

Urban/Global: Politics, Policy, and Health

It is an axiom of contemporary urban studies scholarship that cities can only be understood in terms of both their "internal" characteristics and their connections to other scales, places, and processes. This "global" sense of urban place (Massey 1991, 1993, 2005, 2007) applies to all aspects of cities. More specifically, in the context of this book, we can see that an approach that takes seriously the dialectics of territoriality–relationality and fixity–mobility offers a great deal of insight into urban questions. As Ward (chapter 4) puts it, "if it were ever enough to account for change in the nature of urban development on the basis of analysis generated solely from within cities and the countries of which they are part, then that time has surely passed."

A particular understanding of urban politics parallels this global–relational perspective. Harvey's approach to the urban and to the political is evident in the following passage:

> When I speak of urban politics . . . I do not mean the mayor or the city council, though they are one, important form of expression of urban politics. Nor do I necessarily refer to an exclusively defined urban region, because metropolitan regions overlap and interpenetrate when it comes to the important processes at work there . . . To the degree that the processes are restlessly in motion, so the urban space is itself perpetually in flux. (Harvey 1989, 127)

The resonance of this argument can be felt in statements by other key figures in urban political geography. For example, Cox (2001, 756) argues that "what is commonly referred to as 'urban politics' is typically quite heterogeneous and by no means referable to struggles within, or among, the agents structured by some set of social relations corresponding unambiguously to the urban."

This conceptualization of urban politics must be applied to and developed through the study of concrete cases. The case of policy transfer among cities and the political struggles that are interwoven with it is an ideal opportunity to develop analyses of urban politics. Policy actors, broadly conceived to include institutions and individuals within the formal structures of the state, a range of private policy consultants, academics, and activist groups, are continually looking elsewhere to identify, learn about, and, in some cases, adopt "best" policy practices (Wolman 1992; Dolowitz and Marsh 2000; Theodore and Peck 2000; Wolman and Page 2000, 2002; Stone 1999; Peck and Theodore 2001, 2009; Peck 2003, 2006; McCann 2008, 2011; Hoyt 2006; Ward 2006; Cook 2008). These come in the form of formally drafted guidelines for governance (policies), statements of ideal policies (policy models), or expertise and know-how about good policymaking and implementation (policy knowledge). The politics of policymaking, where various interests struggle over the character and implications of specific sets of guidelines and visions is, as I will illustrate and as the other authors in this volume discuss, generally about more than the city. It entails discussions that often range globally, as different local interests interpret and debate the character, outcomes, and local appropriateness of policies developed and implemented at various points of reference elsewhere.

The literature on policy transfer in political science (Dolowitz and Marsh 2000; Stone 1999, 2004; Evans and Davies 1999; Radaelli 2000; James and Lodge 2003; Evans 2004) offers some valuable insights into the range of institutions and actors who transfer policies. It also specifies a range of different types of transfer (voluntary, coerced, etc.), and it sheds light on the conditions that initiate transfers and that determine their success. However, the literature is limited in three ways: (1) it is unduly bound by narrow typologies of "transfer agents"; (2) it tends to only see transfers happening at the national and international scales, ignoring interactions among cities in different countries; and (3) it displays what Peck and Theodore (2001, 449) call an "implicit literalism" in its definition of policy transfer that tends to assume that policies are transferred from one place to another relatively intact while ignoring the modifications and struggles that occur along the way. It is often an asocial, aspatial, and, ironically, somewhat apolitical literature (for a full critique, see McCann 2011). Recent work by geographers and others has sought to overcome these limits and push further on the complex spatialities and power relations of policy transfer (Peck 2003; Ward 2006; Cook 2008; McCann 2008; chapter 4). For example, the notion of "policy mobilities" draws attention to the social and political character of policies, policy models, and policy knowledge as they are produced, translated, transformed, and deployed by various actors in a range of contexts. A key element of this new work is on cities. It has primarily entailed studies of elites who mobilize best practices to foster urban "livability" and "creativity" (Peck 2005; chapter 3), create Business Improvement Districts (Hoyt 2006; Ward 2006, 2007; Cook 2008; chapter 4), or shape new urban forms and associated services (McCann 2011; and chapters 1, 2, 6, and 7).

Yet there are at least three related themes that the geographical literature on policy mobilities, or policies in motion, might advance further. First, scholarship on *urban* policy mobilities has only recently emerged and has, understandably, focused on detailing and conceptualizing the actors, mechanisms, and contexts through which policies are mobilized. Therefore, there is a great deal of scope both for further analyses of these transfer processes and for analysis of the local *political* consequences of policy mobilities—how transfers stem from and shape urban power relations and political struggles. Second, the literature has so far employed a limited, although very useful, set of examples and case studies, as I note earlier. There is scope for broadening the range of examples used to analyze interurban policy mobilities beyond those that primarily address urban economic development. The study of policies aimed at governing other, not unrelated, aspects of urban life, including environmental concerns, public health, and urban cultures, can inform and benefit from the policy mobilities perspective (see chapters 3 and 6). Furthermore, case studies that address conditions beyond the richest countries are also necessary (e.g., chapters 1 and 2), as are those that seek to historicize contemporary rounds of policy mobility (chapter 3). Third, the existing literature has largely addressed the role of elites—actors within the state at various scales, business coalitions, professional organizations, transnational institutions, think tanks, and consultants—in shaping policies and setting them in motion across the globe. This work is necessary, yet it might be built on and extended by the study of how nonelites, or "subaltern" groups, and social movements inhabit and redirect existing global informational infrastructures and circuits of knowledge or create their own sites and circuits of persuasion to upturn established policies and mobilize alternatives (Bosco 2001; McCann 2008).

The case of political struggles over urban drug policy provides the opportunity to develop these themes within urban geography. Indeed, DeVerteuil and Wilton (2009) note that geographers' engagement with what they call "geographies of intoxicants" has been limited, while Brown (2009) and Ali and Keil (2006, 2007, 2008; Keil and Ali 2007; chapter 6), argue that "traditional" concerns in urban geography would benefit from a more sustained attention to public health. Ali and Keil (2007, 847), drawing on the notion of urban health governance, argue that

> rather than operating solely in between the often contradictory challenges of social cohesion and economic competitiveness, urban governance may soon have to be more centrally concerned with questions of widespread disease, life and death.

Van Wagner (2008, 19) argues that a sharpened focus on urban health allows further refinement of our conceptualization of urban–global relations and suggests that in terms of health, "certain cities emerge as disproportionately influential and [globally] connected," a point that she sees as "important and [that] should be expanded upon." Certainly, Vancouver is connected to a global archipelago of cities with similar harm reduction approaches. These cities and their connections with regional and national governments, global institutions such as the World Health Organization (WHO) and the Joint United Nations Programme on HIV/AIDS (UNAIDS), and organizations like IHRA constitute a network through which best practices are mobilized to address the health, social, and economic harms of illicit drug use. These circuits, mobilities, and flows are usually shaped by territorial configurations and legacies that produce uneven landscapes of health regulations, funding regimes, and political opportunity structures, which, in turn, become objects of and tools for political struggle. In the following pages, I will outline the relational–territorial elements of the politics of drug policy in Vancouver both at the time of the transfer of harm reduction into the city and in the years since.

Looking for a Fix: Finding a Solution to Vancouver's Drug-Related Health Crisis

The search for appropriate and effective approaches to the harmful use of drugs has been a major political issue in Vancouver since the mid-1990s. In 1994, a report from the then chief coroner of British Columbia identified "an epidemic of illicit drug deaths" in the province over the previous six years, marked by an 800 percent increase in fatal overdoses of heroin or cocaine (Cain 1994, 6). The report also noted that 60 percent of these cases occurred in Vancouver, a city that, at the time, contained slightly less than 14 percent of the total provincial population (BC Stats, n.d.). It also critiqued "the so-called 'War on Drugs' . . . as an expensive failure" (Cain 1994, vi), and advocated instead for serious study of the decriminalization and legalization of certain drugs and for strengthening the harm reduction approach that had, since 1987, been the framework for the Canadian government's National Drug Strategy.

Despite this report, Vancouver's public health crisis continued apace through the 1990s. Some 1,200 overdose deaths were recorded from 1992 to 2000 (Wood and Kerr 2006a), and its intravenous drug users suffered extremely high rates of life-threatening bloodborne infection, specifically hepatitis C and HIV, with the annual incidence rate of the latter peaking at 18 percent in 1997—the highest rate ever recorded among an intravenous drug user (IDU) population in the developed world (Wood and Kerr 2006a). The epicenter of the crisis was in the streets, alleys, and single-room occupancy hotels of the Downtown Eastside. The neighborhood, widely regarded as Canada's poorest (Eby and Misura 2006), was the site of an open drug scene, and has long been home to a concentrated service-dependent population, including many homeless or marginally housed people with concurrent addiction and mental health diagnoses. The dire conditions on the Downtown Eastside encouraged public discussion of drug policy, and reactions to the crisis ranged widely and changed markedly in the 1990s, culminating in strong public support for the new harm reduction policy (Wood and Kerr 2006b).

This sea change in public discourse was the result of hard political work by an informal but strong coalition that included a user-run, nonprofit support and advocacy group, a group of parents of drug users, some politicians and officials at all levels of government, social services agencies and nongovernmental organizations (NGOs), researchers, and members of the local police force. They shared an interest in changing how drug use was governed in Vancouver. This coalition exerted political pressure on government at all scales to allow the four pillar approach to be instituted in 2001 (MacPherson 2001). Intertwined with the coalition's political activism at home was a global search for a new policy model that entailed identifying exemplary cases of alternative drug policy, specifically in Swiss cities and in Frankfurt, Germany. The purpose was to educate as many local decision makers as possible about the benefits and challenges of adopting a similar approach (McCann 2008).

Largely because of this work, harm reduction has become a central part of the discourse and practice of drug policy in Vancouver, and opposition seems to have waned in the wake of the apparent benefits of the approach, including a decrease in drug-related deaths (Matas 2008) and, more specifically, the demonstrated benefits of Insite, including overdose prevention, counseling, detox, and treatment referrals, reductions in syringe sharing, public injections and public disposal of syringes (Wood et al. 2006; Urban Health Research Initiative 2009), and the local HAT trial (NAOMI 2008a, 2008b). Nonetheless, skeptics and opponents continue to present their case against harm reduction not only in terms of their own interpretations of its operation in Vancouver but also in terms of their understandings of the merits of drug strategies elsewhere—understandings that have been supported by opponents' own fact-finding trips to such places as Frankfurt and San Patrignano. The scheduling of the San Patrignano meeting on the same night as the IHRA-related forum underscores this ongoing debate. Vancouver's drug policy was and continues to be understood and debated *in terms of* other points of reference.

Re: Frankfurt and Zürich

As the harm reduction coalition grew in the 1990s, a first goal was to identify a feasible alternative model and a real-world example of its implementation. It was clear that the choice of a point of reference was crucial not only in policy terms but also in political terms. The exemplar had to be able to be made understandable to decision makers and the public:

So [we] . . . started saying, "Well, who can we learn from? Where are the politics most similar? Where, from all these different things we've heard about from different parts of the world, where do we need to go and learn?" And we really concluded Frankfurt was the spot. That had to do with the federal structure of German government . . . [and] it was a city that wasn't too big . . . [W]e thought that people [in Vancouver] could kind of get their heads around this city in Germany perhaps a little better than some of the other places that were doing supervised injection at the time. So that's how we picked Frankfurt. (interview, NGO representative no. 1, 2006)

One key element of this choice was the similar organization of each federal state, both of which has municipal, provincial/länder, and national tiers. This, the coalition believed, would allow common issues of overlapping jurisdictions and multiple service providers and regulatory agencies to be addressed. However, they also acknowledged that the division of powers and funding among the three tiers differed from Germany to Canada—a point that, as I will discuss later, others subsequently identified as a problem with the pro–harm reduction argument.

If a harm reduction model was to be transferred from Europe, the activists also decided that bureaucratic reports would not be enough to win the inevitable political debate. To "debunk anxieties and concerns [and] closed thinking in the bureaucracy" (interview, NGO representative no. 1, 2006), it was necessary to make the Frankfurt model more tangible. Two main strategies emerged: (1) find ways to have as many key players as possible visit Frankfurt and (2) find ways to have key figures from Frankfurt visit Vancouver:

> I think that when you tell people that you've actually seen it, they lend greater credence to what you're saying because, before that, well, one of the main questions is, "Well, have you ever seen one [a supervised injection site or a prescription heroin facility]? Have you ever been there?" . . . Personal experience cannot ever be underestimated, right? And I don't necessarily think it must mean that we go there. Sometimes people [from] Frankfurt could come here and say, "Oh you know this is exactly what's going on." . . . You know, it normalizes it. (interview, social service agency representative, 2007)

The coalition organized a visit to Frankfurt and included a journalist, a documentary filmmaker, and a Vancouver police drug squad officer in the delegation. After one week, the group returned largely impressed and convinced by their engagement with Frankfurt. Beyond the firsthand stories that they circulated informally upon their return, the trip also produced a positive report from the police officer, material for in-depth articles by the journalist, and footage for a documentary film, which appeared in the public sphere just before the city council was to vote on the adoption of its new drug strategy. On the Vancouver side of the relationship, a local social services society organized a conference in which European experts explained their policies to an audience gathered under a tent in a Downtown Eastside park. This suggests that fact-finding trips and consultants' visits are physical, embodied activities that are valuable in policy transfer processes because they add substance to the types of information gleaned from the Internet and policy reports. They are not only about traveling to learn or to teach, however. The physical, rather than virtual, experience of place and of process is a powerful element of political persuasion, lending credence to arguments about exemplars elsewhere through visual documentation, personal anecdotal accounts, face-to-face, and peer-to-peer contacts (McCann 2011).

Skeptics' Reports

Of course, all actors in a political debate can invoke points of reference from elsewhere and offer different experiences and interpretations. Those who questioned the positive narrative of European harm reduction strategies expressed their own interpretations in the political debate leading up to the decision on the new drug policy. First, as is frequently the case in the politics of policy transfer, concerns were raised about the fit between policy models from elsewhere and the local context. For example, a city councilor was funded to visit drug programs in Amsterdam and Frankfurt in 2000. Her report to council (Clarke 2000), while by no means dismissive of the merits of these cities' approaches, was lukewarm in comparison to the report delivered the previous year by the planner. "To understand the context for what these two cities have been able to do," she argued, "it's important to know they each have the legal jurisdiction and budget for what we in Vancouver would see as a combination of provincial and city functions." Even when the new drug strategy had been approved by the majority of councilors, skeptics pointed out that, as the *Vancouver Sun*'s editors summarized it, "the city only has the power to act on four of the approximately 36 points in the proposal and that other agencies—including the provincial government—will have to come on board for it to be successful" (*Vancouver Sun* 2001).

A second specter that haunted pro–harm reduction forces was Zürich's "Needle Park." While the coalition had focused primarily on Frankfurt, it had continued to invoke Swiss policy as another example of good harm reduction policy. Opponents, however, noted that Needle Park was a failed experiment in urban harm reduction and argued that it was evidence of why Vancouver should not move in this direction. In 1987, as the global HIV epidemic grew, Zürich's authorities decided that it was best to control and concentrate illicit drug use in one location to reduce its impact on the entire city and to best provide services to users. They forced the city's formerly scattered drug scenes to congregate in Platzspitz, a central-city park. The use and sale of drugs was tolerated within the park and a range of social and health services were provided, including needle and syringe exchange, resuscitation equipment, counseling, employment services, shelter, food, toilets, and bathing facilities.

Although the initiative allowed greater contact between hard-to-access users and service providers, a "honeypot" effect emerged, in which an increasing proportion of the users came from other parts of Switzerland or from abroad. On the supply side, the park became the focus of increasingly violent competition among organized dealers vying for shares of a captive market. As the park grew more violent, it became less feasible for service providers to work in it, thus compromising one of its main raisons d'être. This combined with the increasingly large numbers of people injecting in the open and the litter-strewn nature of the space led to the media's use of the sobriquet, "Needle Park," a nickname that crystallized increasing neighborhood opposition to its existence. The park was closed in 1992, spurring fears among service providers that many users would again become hard to access and would therefore be more vulnerable to HIV infection, overdose deaths, and other drug-related harms (*Province* 1992; Grob 1993; Huber 1994; Foulkes 2002).

Stories of Needle Park made their way to Vancouver in the early 1990s, through national and local media sources (Drohan 1991; *Province* 1992). A survey of newspapers shows that there was a subsequent lull in references to the Zürich situation, locally and nationally, until 1997, when Vancouver's health authorities declared the health emergency. At this point, there was an uptick in references to the park, both in news reporting and in letters to the editor. The push for harm reduction in Vancouver, which drew from Switzerland's *post*–Needle Park experience (MacPherson 1999, 2001), nonetheless raised fears of a "Needle Park on the Pacific" (Diewert 1998), with a similar honeypot effect and related increases in drug-related violence and litter in the Downtown Eastside and its surrounding neighborhoods. As one key proponent argued, skeptics would "point to Zürich. But they'd have their facts wrong. Because they'd heard that Zürich did something and then they'd say, 'That was a total disaster. Look at "Needle Park."' And then we'd say, 'Well, no, "Needle Park" was before'" (interview, drug policy official, 2005). Nevertheless, it is perhaps no surprise that a key element of the pro–harm reduction narrative in Vancouver involved a commitment to public order, to dealing with the open drug scene in the Downtown Eastside, and to reducing the harms caused to neighborhoods by carelessly discarded needles and other risky behaviors.[4]

Switzerland was also a reference point for a third critique of Vancouver's proposed drug policy. Advocates' idea of opening a HAT clinic was intended to address the difficulties faced by chronic heroin users, for whom abstinence-oriented treatment and methadone prescription had persistently failed. They were inspired by what they saw as the promising results of a similar program in Switzerland and by plans to replicate it in Germany (MacPherson 1999). The Swiss trials began in the aftermath of Needle Park and early results, which had begun to be reported before Vancouver's coalition looked to Swiss cities for lessons, suggested that when users were assured a heroin supply that was legal and safe (i.e., of a consistent, known dosage with no adulterants, unlike illicit heroin), they would be healthier, would be more likely to enter treatment programs, including abstinence-based programs, and would be less involved in crime and more likely to maintain housing and legal employment

(Fischer and Rehm 1997; Uchtenhagen 1997; Marlatt 1998; Drucker 2001; Rehm et al. 2001).

Yet opponents repeated critiques of the methodology of the trial that were initially laid out in an otherwise cautiously optimistic evaluation conducted by a WHO panel (Ali et al. 1999). They noted that it had not used a randomized control test methodology (in which participants would be randomly and blindly assigned to a group taking prescription heroin or to a control group taking a substitute, such as methadone) and that users were asked to self-report their levels of use. For these and other methodological and contextual reasons, opponents contended that HAT was unproven and should not be given credence or resources in Vancouver (Satel 1998; Lawson 1999), even though the WHO report suggested that more research trials in other places were exactly what was necessary.

Points of Reference and the Debate in Vancouver

Vancouver's harm reduction advocates were suspicious of their critics:

> The naysayers were looking at Europe and they were saying "Yeah, they've got all this [research] but it wasn't sufficiently rigorous . . ." So . . . [the politician who visited Frankfurt and Amsterdam] was standing there and dissing Switzerland's prescription heroin project because . . . she could pick holes in the scientific methodology . . . And so, the people who did not want to go in this direction would use that as the crutch to halt everything. (interview, NGO representative no. 2, 2007)

This indicates that the debate, in part, involved a struggle to define the parameters of comparison and success that would guide the development of a "Vancouver model" of harm reduction in reference both to local context and to global precedents. The extent of harm reduction's success in Frankfurt had to be understood in terms of the structure and capacities of the Canadian state.

Proponents of harm reduction did not necessarily disagree with this point. They were, for example, intent on properly and sensitively embedding the general principles of harm reduction in the local context, or as one key member of the coalition remembered, "At one point I finally said . . . 'I don't need a Made in Frankfurt solution, I need a Made in the Downtown Eastside solution.' And we did it" (interview, user-activist, 2007). Yet the coalition remained wary of the intentions of those skeptics who seemed to voice support for drug policy innovation in principle but who were, they felt, also looking for any excuse to drag their feet. Referring to the councilor's report from Europe, a coalition member argued the following:

> She . . . went over, and any little sort of whisper or [anything she found that was] not quite exactly perfect or [had] a little bit of debate about [it], her mind would pick that out and bring it back . . . She'd be the "yes, but" person. And she would always sound like she supported [harm reduction], but in the end it was always "later" . . . So, I think it came down to just morally she couldn't go there. (interview, advocacy organization representative, 2006)

Questions of morality, evidence, and credibility continue to mark the politics of drug policy in Vancouver, as continued references to Needle Park (e.g., Chua 2006; *Province* 2006) and the Swiss heroin trials (e.g., Sabet 2005; Kendall 2005; McKnight 2006) in local media attest. This underscores my argument that reference points of debate deployed and contested at the time of a particular policy transfer continue to resonate in and frame the parameters of discussion long after policies from elsewhere are territorialized in a new location.

Re: San Patrignano

It is not only that specific examples continue to be debated over time, however, but also that new points of reference are invoked as debate continues to bolster or to question orthodoxies about how policy should be enacted. San Patrignano, the subject of the smaller of the two public meetings held in Vancouver in May 2006, exemplifies this use of reference points in the politics of urban policy. A survey of Canadian newspapers identifies only three relevant references to the rural Italian abstinence-oriented therapeutic community before September 2006. There have been more than forty since that date, however. The year 2006 marked the public rollout of San Patrignano as a model alternative or complement to the four pillars, Insite, and NAOMI. Its proponents express skepticism about the benefits of harm reduction and a belief that more needs to be done to treat people with drug addictions, rather than maintaining their addictions through the prescription of methadone and heroin or through supervised consumption. Some see the four pillars approach and particularly Insite as a failure while others see San Patrignano as complementary to harm reduction. One powerful, well-connected advocate of the Italian model (not the person who held the meeting in May 2006) argues that Insite should be closed because it has not worked and is a waste of money (an argument that runs counter to the scientific evidence produced by the medical researchers contracted to evaluate the facility (Wood et al. 2006):

> The reason that we started the Supervised Injection Site—and I was one of the people that advocated for it, I had to go and get money for it from my government . . . was [to] . . . reduce street disorder. We thought we would reduce the spread of HIV and Hepatitis C and we thought we would have more people going into detox and to treatment . . . And none of those things have happened. You know, street disorder has never been worse. We have Hepatitis C in 90% of the addicts in the downtown core . . . So we haven't affected a change there . . . [W]e've tried it and it didn't work. (interview, provincial politician, 2007)

Politically, the debate involves a distinct turn of the tables. Whereas in the 1990s and early 2000s harm reduction advocates saw themselves as battling against feet-dragging and downright hostile opponents, by the mid-2000s harm reduction had become "the establishment" in Vancouver, if not elsewhere in Canada. Thus, local critics of harm reduction accuse the harm reduction practitioners of using the powers of the state to drive through their agenda while closing out other opinions. This accusation presents a particular political challenge for harm reduction practitioners and advocates who take seriously the movement's philosophy of non-judgmental pragmatism that encourages an acceptance of any model that might make a difference. Vancouver's drug policy coordinator referred to a proposal to operate a San Patrignano–inspired therapeutic community in rural northern British Columbia as "a compelling idea," which "would be a welcome addition to the array of options that we have for people" (in Bermingham 2007). Furthermore, his office has recently responded positively to the opening of another therapeutic community for youth in British Columbia (City of Vancouver 2009).

Nonetheless, San Patrignano tends to be presented as a replacement or corrective to harm reduction by its leading proponents, thus casting the politics of drug policy in a contentious light, which, as I will discuss later, is reflected somewhat in the responses of some harm reduction proponents. First, however, it is worth noting that proponents of San Patrignano are keenly aware of the political power of using a long-standing model from elsewhere to support their position, just as harm reduction proponents understood the power of invoking German and Swiss examples a decade ago. When asked about the benefit of being able to point to a case like San Patrignano when advocating for new policy, one proponent echoed the words of those who had previously invoked Frankfurt as a model for Vancouver:

It's very useful. First off, you know, it takes quite a lot to imagine this kind of a model . . . It certainly makes sense once you get it . . . But it's very hard to imagine that you could affect that many people's lives. So it's very important that, you know, as someone that's at the beginning of it, that I have a vision that I can feel comfortable talking about. You know, I've actually seen it. I know it will be different here, but I've seen it. And I can look at their results. (interview, provincial politician, 2007)

Contemporary advocates of San Patrignano also echo the earlier efforts of harm reduction proponents when they acknowledge that only some elements of the Italian approach are suited to Canada and that the model must be modified for its new context. I have already suggested that a key attraction of San Patrignano is its emphasis on treatment and its distaste for harm reduction. A second appeal seems to be San Patrignano's arm's-length relationship with state funding. Whereas proponents of this model tend to critique the Canadian state at all levels for an abundance of red tape, its overfocus on harm reduction, its tendency to fund short-term treatment programs, and a general aversion to innovation, San Patrignano offers a private solution. The community takes no operating funds from the Italian government and relies instead on the sales of commodities—everything from wine and honey to horses and bikes—that are produced on-site by recipients of treatment. This suggests that there may be an ideological as well as a moral attraction to San Patrignano among some of its proponents. It offers a vision of drug treatment through the private rather than the public sector. Furthermore, the trades-based instruction and production that characterize the community appeal to many proponents. One visitor to San Patrignano in particular, who is a skilled artisan and who has run programs to train at-risk youth, was particularly impressed by this aspect of the community.

Proponents of San Patrignano are less comfortable with some of its defining, world-renowned features, however. For example, they find it difficult to embrace San Patrignano's approach to drugs. On the one hand, alcohol (specifically wine, which is viewed as a digestive and served with meals) and tobacco are both permitted at the community. On the other hand, San Patrignano is founded on an aggressive, frequently repeated abhorrence of illicit and medically prescribed drugs and their use. One Vancouver advocate is skeptical of this approach and its implication that methadone cannot be used as part of treatment:

> I think there is great merit for treatments where people are being weaned
> off of drugs. So for example, somebody is a heroin addict and they go

on methadone and they are slowly coming off of it. That seems like
a legitimate recovery effort . . . Whereas in San Pat, it might be completely, "You wouldn't do it." (interview, provincial politician, 2007)

The discomfort expressed by proponents of the San Patrignano model
suggests again that one of the main reasons for identifying a policy exemplar from elsewhere as a model of how things might be done at home
is strategic and political. It is likely that as the San Patrignano model is
operationalized in British Columbia, it will resemble its ancestor in only
some ways and may resemble other forms of therapeutic community more
closely. Yet the ability to crystallize a political position—one critical of
harm reduction and supportive of private, long-term, abstinence-oriented
residential treatment—through the shorthand of a model from elsewhere
is strategically attractive.

Certainly, it has encouraged various responses in Vancouver. While
some harm reduction advocates and practitioners have cautiously welcomed the approach as a complement to existing strategies, others question
the model, given the history of abuse allegations at the community and
the open questions surrounding its success rate. They are suspicious of
the assumptions and intent underlying local San Patrignano advocacy. "I
know all about San Patrignano and it's not everything that it's cracked
up to be either," says one activist (interview, user-activist, 2007). Another
member of Vancouver's harm reduction coalition argues that

> the San Pat model . . . if you try to disseminate it into here, it's not very
> practical . . . It's very expensive and people live there for an awfully long
> time. And it's kind of a separate isolated community . . . It feels like it's
> creating [an] unreal world to replace another unreal one . . . It doesn't
> seem to me to be a long-term sustainable solution. Maybe I love it as
> an interim . . . (interview, advocacy organization representative, 2006)

Yet for her, the turn among some in Vancouver to the San Patrignano
model is another example of intractable differences in belief, where proponents say, "Okay, let's pick that idea that fits into their belief system"
(interview, advocacy organization representative, 2006).

San Patrignano, like Frankfurt and Zürich, remains a point of reference
in Vancouver's politics of drug policy. Indeed, one of the plans for a San
Patrignano–inspired therapeutic community, although not the one promoted at the meeting in May 2006, has now come to fruition. New Hope
operates on a former U.S. military radar station in a rural area outside the
northern British Columbia city of Prince George, a 450-mile drive from
Vancouver (Bermingham 2007). The community currently houses one

hundred residents, half of whom are from Vancouver. In Vancouver, harm reduction remains a central focus of drug policy and has shown significant success. NAOMI's first phase has ended, leaving in question the futures of the city's chronic opiate users, and the future of Insite is continually under threat from an unsympathetic federal government. Debate over the practice and outcomes of harm reduction continues in the city, and those involved seldom miss the opportunity to refer to places elsewhere.

Conclusion

In their discussion of the urban regions as political assemblages, where various forces that might generally be viewed as existing elsewhere or at other scales are seen as assembled in the urban region for the purposes of governance, Allen and Cochrane (2007, 1171) argue that,

> increasingly, it would seem that there is little to be gained by talking about regional governance as a territorial arrangement when a number of the political elements assembled . . . are "parts" of elsewhere, representatives of professional authority, expertise, skills and interests drawn together to move forward varied agendas and programmes . . . There is . . . an interplay of forces where a range of actors mobilize, enrol, translate, channel, broker and bridge in ways that make different kinds of government possible.

This is not to say that the city, or urban politics, does not have materiality or powerful consequences. Of course they do. What Allen and Cochrane, as well as many other scholars of urban politics and policymaking I refer to in this chapter, acknowledge is that urban politics is always about more than the city, both in terms of its consequences and in terms of its referents.

Vancouver's politics of drug policy is constituted by the very real and very local concerns of the Downtown Eastside, but it is also shaped by travels to, stories from, and relations among a range of other places. The city's drug policy and the politics that surround it are studded with these "parts of elsewhere" and are, therefore, both territorial and global–relational assemblages. Nonetheless, Allen and Cochrane, among others, remind us that it is not enough for us to acknowledge the global–relational character of urban policy by providing detailed accounts of local or territorial politics and then simply gesturing "up" to the wider global context as "obviously" playing some constitutive role in the local process. Equally problematic is a focus on global relations among cities that then gestures "down" to quickly sketched examples from specific cities or territories to bolster or validate the global analysis. Contemporary literatures on scale, global cities, and urban neoliberalism have moved beyond the allures of "gestural analysis" toward well-conceptualized but also empirically detailed investigation of the global and the local as they are combined in certain moments, by and for certain interests (Burawoy et al. 1991, 2000).

My discussion of the deployment of evidence from elsewhere in the politics of drug policy in Vancouver is an attempt at such an empirical approach that holds in its sights a balance between the place-based politics of one city and the global relations that constitute a global network of cities with similar approaches to drug policy. More specifically, my discussion of the ways in which particular discursive constructions of Frankfurt, Zürich, and San Patrignano get put to work in Vancouver's public sphere speaks to the burgeoning literature on policy transfer, policies in motion, and policy mobilities. There is room in this literature for more detailed qualitative investigations of how the adoption and operationalization of policies, policy models, and policy knowledge from elsewhere shapes urban politics. While the literature has begun to engage this question in a time frame usually focused on the lead-up and immediate aftermath of a new policy's importation, I have argued that the specific exemplars from elsewhere seem to linger long after a policy has been adopted and remolded into a local solution. The Vancouver example suggests they can resonate for a decade or more. The specific time horizon will likely vary depending on context and presumably will diminish over time—a temporal version of distance decay. Nonetheless, these parts of elsewhere remain as frames, referents, and points of contention in future policy debates.

Moreover, I also argue it is not only the *content* of a particular policy debate that resonates on into the future but also a particular *form* of argumentation and practice becomes "lodged," as Allen and Cochrane put it, in the public sphere. I have found no evidence that the San Patrignano model was a matter of serious public discussion at the time of Vancouver's search for a new model of drug policy at the end of the 1990s. Yet since 2006, it has become a frequently discussed complement, corrective, or replacement for harm reduction. I would suggest that the utility of San Patrignano as a political counter to establishment harm reduction in Vancouver is in no small part the result of previous rounds of conditioning in the public sphere where participants in policy debates have become used to and might even expect to be persuaded of the merits of a new policy proposal through references to evidence from elsewhere. In this regard, it is worthwhile, taking a long view of the politics of policy transfer, to see it as both relational and territorial but also short term and long term.

Notes

The research presented here was funded by a Social Sciences and Humanities Research Council Standard Grant and is part of a larger project on the development of Vancouver's drug policy. I am extremely grateful to those who agreed to be interviewed. Thanks also to Stephanie Campbell, Nicole Kennedy, Lynn Saffery, Rini Sumartojo, and Cristina Temenos for research assistance and to Kevin Ward for comments on an earlier draft. All errors of fact and interpretation are mine.

1. Australia, Canada, Germany, Luxembourg, The Netherlands, Norway, Spain, and Switzerland (Hedrich 2004).

2. Not all in the global harm reduction movement would agree with this goal.

3. These were not the only reference points, although they became the most important. Others included Amsterdam (which has a long history of and global reputation for a nonpunitive approach to drug use), Merseyside, United Kingdom (where the regional health authority and police force were early proponents of harm reduction in the 1980s), and Portland, Oregon (where the Central City Concern organization has, since the 1970s, developed a continuum of care for marginalized, homeless people who are addicted to alcohol and other drugs).

4. Proponents also noted that the problems that occurred at Needle Park were not an indictment of harm reduction in toto. As a Swiss commentator puts it, "we can learn from the Zürich experiment that tolerating an open drug scene can have fateful consequences, especially when combined with extensive measures of harm reduction. It is not the policy of harm reduction that is questioned, but the policy of tolerating an open drug scene" (Huber 1994, 515).

References

Ali, R., M. Auriacombe, M. Casas, L. Cottler, M. Farrell, D. Kleiber, A. Kreuzer, et al. 1999. *Report of the External Panel on the Evaluation of the Swiss Scientific Studies of Medically Prescribed Narcotics to Drug Addicts*. Geneva, Switzerland: World Health Organization. http://www.druglibrary.org/Schaffer/Library/studies/OVERALLS.htm.

Ali, S. H., and R. Keil. 2006. "Global Cities and the Spread of Infectious Disease: The Case of Severe Acute Respiratory Syndrome (SARS) in Toronto, Canada." *Urban Studies* 43, no. 3: 491–509.

———. 2007. "Contagious Cities." *Geography Compass* 1, no. 5: 1207–26.

———, eds. 2008. *Networked Disease: Emerging Infections in the Global City*. Malden, Mass.: Wiley-Blackwell.

Allen, J., and A. Cochrane. 2007. "Beyond the Territorial Fix: Regional Assemblages, Politics, and Power." *Regional Studies* 41:1161–75.

Arnao, G. n.d. "Drug Policy and Ideology: An Italian Case." http://www.drugtext.org/index.php/en/articles/112—drug-policy-and-ideology.

BC Stats. n.d. "BC Municipal Population Estimates, 1986–1996." http://www.bc
stats.gov.bc.ca/DATA/pop/pop/mun/Mun8696a.asp.

Bermingham, J. 2007. "Centre Near Prince George Could Offer New Hope for
Addicts." *The Province,* June 26.

Bosco, F. J. 2001. "Place, Space, Networks, and the Sustainability of Collective
Action: The Madres de Plaza de Mayo. *Global Networks* 1, no. 4: 307–29.

Brown, M. 2009. "Public Health as Urban Politics, Urban Geography: Venereal
Biopower in Seattle, 1943–1983." *Urban Geography* 30, no. 1: 1–29.

Burawoy, M., J. A. Blum, S. George, Z. Gille, T. Gowan, L. Haney, M. Klawiter,
et al. 2000. *Global Ethnography: Forces, Connections, and Imaginations in a
Postmodern World.* Berkeley: University of California Press.

Burawoy, M., A. Burton, A. A. Ferguson, K. J. Fox, J. Gamson, L. Hurst, N. G.
Julius, et al. 1991. *Ethnography Unbound: Power and Resistance in the Modern Metropolis.* Berkeley: University of California Press.

Cain, V. 1994. *Report on the Task Force into Illicit Narcotic Overdose Deaths
in BC.* Victoria: Government of British Columbia, Ministry of the Attorney
General.

Chua, T. T. 2006. "Drug Addiction: Disease of Choice." *Coquitlam Now,* September 20, 15.

City of Vancouver. 2009. "Long-Term Youth Residential Treatment Comes to
BC." Vancouver: City of Vancouver Four Pillars Drug Strategy. Podcast audio.
http://vancouver.ca/fourpillars/podcasts/media/TheCrossing.mp3.

Clarke, J. 2000. *Report on Visit to Drug Programs in Amsterdam, Netherlands
and Frankfurt, Germany.* City of Vancouver, September. http://vancouver.ca/
ctyclerk/cclerk/000926/drugs.htm.

Cook, I. R. 2008. "Mobilising Urban Policies: The Policy Transfer of U.S. Business Improvement Districts to England and Wales." *Urban Studies* 45, no. 4:
773–95.

Cox, K. R. 2001. "Territoriality, Politics, and 'The Urban.'" *Political Geography*
20, no. 6: 745–62.

DeVerteuil, G., and R. Wilton. 2009. "The Geographies of Intoxicants: From Production and Consumption to Regulation, Treatment, and Prevention." *Geography Compass* 3, no. 1: 478–94.

Diewert, T. 1998. "Treat Addicts, Don't Provide a Needle Park on the Pacific."
Vancouver Sun, September 21, A11.

Dolowitz, D., and D. Marsh. 2000. "Learning from Abroad: The Role of Policy
Transfer in Contemporary Policy-Making." *Governance* 13:5–24.

Drohan, M. 1991. "Needle Park Gives Zürich a Headache." *Globe and Mail,* November 7, A14.

Drucker, E. 2001. "Injectable Heroin Substitution Treatment for Opioid Dependency." *The Lancet* 358:1385.

Eby, D., and C. Misura. 2006. *Cracks in the Foundation: Solving the Housing Crisis in Canada's Poorest Neighbourhood.* Vancouver, B.C.: Pivot Legal Society.

Evans, M., ed. 2004. *Policy Transfer in Global Perspective.* Aldershot: Ashgate.

Evans, M., and J. S. Davies. 1999. "Understanding Policy Transfer: A Multi-Level, Multi-Disciplinary Perspective." *Public Administration* 77, no. 2: 361–85.

Fischer, B., and J. Rehm. 1997. "The Case for a Heroin Substitution Trial in Canada." *Canadian Journal of Public Health* 88:367–70.

Foulkes, I. 2002. "Ten Years On from Needle Park." http://www.swissinfo.org/eng/front/detail/Ten_years_on_from_Needle_Park.html?siteSect=105&sid=1008665&cKey=1012813920000.

Grob, P. J. 1993. "The Needle Park in Zürich: The Story and the Lessons to Be Learned." *European Journal on Criminal Policy and Research* 1, no. 2: 48–60.

Haden, M. 2004. "Regulation of Illegal Drugs: And Exploration of Public Health Tools." *International Journal of Drug Policy* 15:225–30.

Harvey, D. 1989. *The Urban Experience.* Baltimore: Johns Hopkins University Press.

Hedrich, D. 2004. *European Report on Drug Consumption Rooms.* Luxembourg: European Monitoring Centre for Drugs and Drug Addiction/Office for Official Publications of the European Communities.

Hoyt, L. 2006. "Imported Ideas: The Transnational Transfer of Urban Revitalization Policy." *International Journal of Public Administration* 29, nos. 1–3: 221–43.

Huber, C. 1994. "Needle Park: What Can We Learn from the Zürich Experience?" *Addiction* 89:513–16.

James, O., and M. Lodge. 2003. "The Limitations of Policy Transfer and Lesson Drawing for Public Policy Research." *Political Studies Review* 1, no. 2: 179–93.

Keil, R., and S. H. Ali. 2007. "Governing the Sick City: Urban Governance in the Age of Emerging Infectious Disease." *Antipode* 39, no. 5: 846–73.

Kendall, P. 2005. "'Harm Reduction' Works in Europe." *Vancouver Sun*, March 22, A13.

Lawson, B. 1999. "Switzerland's Heroin Program Faces the Voters: Proponents Say Providing the Drug to Addicts Is Working; Opponents Call It 'A Subterfuge to Legalize Drugs.'" *Globe and Mail*, June 9, A19.

MacPherson, D. 1999. *Comprehensive Systems of Care for Drug Users in Switzerland and Frankfurt, Germany.* Vancouver, B.C.: City of Vancouver Social Planning Department.

———. 2001. *A Framework for Action: A Four Pillar Approach to Drug Problems in Vancouver.* Vancouver, B.C.: City of Vancouver Office of Drug Policy.

Marlatt, G. A. 1998. "Basic Principles and Strategies of Harm Reduction." In *Harm Reduction: Pragmatic Strategies for Managing High-Risk Behaviours,* edited by G. A. Marlatt, 49–68. New York: Guilford.

Massey, D. 1991. "A Global Sense of Place." *Marxism Today*, June, 24–29.

———. 1993. "Power-Geometry and a Progressive Sense of Place." In *Mapping the Futures: Local Cultures, Global Change*, edited by J. Bird, B. Curtis, T. Putman, G. Robertson, and L. Tickner, 59–69. New York: Routledge.

———. 2005. *For Space*. Thousand Oaks, Calif.: Sage.

———. 2007. *World City*. London: Polity Press.

Matas, R. 2008. "BC Drug Deaths Hit a Low Not Seen in Years." *Globe and Mail*, December 9.

Maté, G. 2008. *In the Realm of Hungry Ghosts: Close Encounters with Addiction*. Toronto, ON: Knopf Canada.

McCann, E. J. 2008. "Expertise, Truth, and Urban Policy Mobilities: Global Circuits of Knowledge in the Development of Vancouver, Canada's 'Four Pillar' Drug Strategy." *Environment and Planning A* 40, no. 4: 885–904.

———. 2011. "Urban Policy Mobilities and Global Circuits of Knowledge: Toward a Research Agenda." *Annals of the Association of American Geographers* 101:107–30.

McKnight, P. 2006. "Give the Addicts Their Drugs." *Vancouver Sun*, April 29, C5.

McMartin, P. 2006. "MLA Wants to Ship Addicts to a Rural Tough-Love Facility. But the Italian Model He Advocates Has Drawn Serious Criticism." *Vancouver Sun*, September 28. http://www2.canada.com/vancouversun/news/westcoastnews/story.html?id=2c68cc8d-2525-4ac7-8e8e-7ea29cc7da32&p=1.

NAOMI [North American Opiate Medication Initiative] Study Team. 2008a. "Reaching the Hardest to Reach—Treating the Hardest-to-Treat: Summary of the Primary Outcomes of the North American Opiate Medication Initiative" (NAOMI). http://vancouver.ca/fourpillars/documents/NAOMIResultsSummary-Oct172008.pdf.

———. 2008b. "Results Show That North America's First Heroin Therapy Study Keeps Patients in Treatment, Improves Their Health, and Reduces Illegal Activity." News release. http://vancouver.ca/fourpillars/documents/NAOMIResultsNewsRelease-Oct172008.pdf.

Peck, J. 2003. "Geography and Public Policy: Mapping the Penal State." *Progress in Human Geography* 27, no. 2: 222–32.

———. 2005. "Struggling with the Creative Class." *International Journal of Urban and Regional Research* 29, no. 4: 740–70.

———. 2006. "Liberating the City: Between New York and New Orleans." *Urban Geography* 27, no. 8: 681–713.

Peck, J., and N. Theodore. 2001. "Exporting Workfare/Importing Welfare-to-Work: Exploring the Politics of Third Way Policy Transfer." *Political Geography* 20:427–60.

———. 2009. "Embedding Policy Mobilities." Working paper, Department of Geography, University of British Columbia.

Province. 1992. "We Need Needle Park: Swiss AIDS Workers." January 17, A38.

————. 2006. "Fewer New Users of 'Loser Drug' Heroin: Free Drugs, Needles Alter Thinking of Youth." June 6, A32.

Radaelli, C. M. 2000. "Policy Transfer in the European Union: Institutional Isomorphism as a Source of Legitimacy." *Governance* 13, no. 1: 25–43.

Rehm, J., P. Gschwend, T. Steffen, F. Gutzwiller, A. Dobler-Mikola, and A. Uchtenhagen. 2001. "Feasibility, Safety, and Efficacy of Injectable Heroin Prescription for Refractory Opioid Addicts: A Follow-Up Study." *The Lancet* 358:1417–20.

Riley, D., and P. O'Hare. 2000. "Harm Reduction: History, Definition, and Practice." In *Harm Reduction: National and International Perspectives*, edited by J. A. Inciardi and L. D. Harrison, 1–26. Thousand Oaks, Calif.: Sage.

Sabet, K. 2005. "Why 'Harm Reduction' Won't Work." *Vancouver Sun*, March 19, C5.

San Patrignano. 2008a. "Inside the Community." http://www.sanpatrignano.org/?q=node/216.

————. 2008b. "Other Problems." http://www.sanpatrignano.org/?q=node/4204.

Satel, S. 1998. "The Swiss and Others Should Acknowledge That Welfare and Harm-Reduction Programs Enable Addiction." *Globe and Mail*, June 13, D9.

Stone, D. 1999. "Learning Lessons and Transferring Policy across Time, Space, and Disciplines." *Politics* 19, no. 1: 51–59.

————. 2004. "Transfer Agents and Global Networks in the 'Transnationalization' of Policy." *Journal of European Public Policy* 11, no. 3: 545–66.

Theodore, N., and J. Peck. 2000. "Searching for Best Practice in Welfare-to-Work: The Means, the Method, and the Message." *Policy and Politics* 29, no. 1: 81–98.

Uchtenhagen, A. 1997. *Summary of the Synthesis Report. Programme for a Medical Prescription of Narcotics: Final Report of the Research Representatives.* Zurich: Institute for Social and Preventive Medicine at the University of Zurich. http://www.drugpolicy.org/library/presumm.cfm.

Urban Health Research Initiative of the BC Centre for Excellence in HIV/AIDS. 2009. *Findings from the Evaluation of Vancouver's Pilot Medically Supervised Safer Injection Facility.* Vancouver: BC Centre for Excellence in HIV/AIDS. http://uhri.cfenet.ubc.ca/images/Documents/insite_report-eng.pdf.

Vancouver Sun. 2001. "Vancouver City Council Approves Drug Strategy." May 16.

Van Wagner, E. 2008. "Toward a Dialectical Understanding of Networked Disease in the Global City: Vulnerability, Connectivity, Topologies." In *Networked Disease: Emerging Infections in the Global City*, edited by S. H. Ali and R. Keil, 13–26. Malden, Mass.: Wiley-Blackwell.

Ward, K. 2006. "'Policies in Motion,' Urban Management and State Restructuring: The Trans-Local Expansion of Business Improvement Districts." *International Journal of Urban and Regional Research* 30:54–75.

————. 2007. "Business Improvement Districts: Policy Origins, Mobile Policies, and Urban Liveability." *Geography Compass* 2:657–72.

Wolman, H. 1992. "Understanding Cross-National Policy Transfers: The Case of Britain and the U.S." *Governance* 5:27–45.

Wolman, H., and E. Page. 2000. *Learning from the Experience of Others: Policy Transfer among Local Regeneration Partnerships*. York: Joseph Rowntree Foundation.

———. 2002. "Policy Transfer among Local Government: An Information-Theory Approach." *Governance* 15, no. 4: 477–501.

Wood, E., and T. Kerr. 2006a. "What Do You Do When You Hit Rock Bottom? Responding to Drugs in the City of Vancouver." *International Journal of Drug Policy* 17:55–60.

———, eds. 2006b. "Cities and Drugs: Responding to Drugs in the City of Vancouver, Canada." Special issue, *International Journal of Drug Policy* 17, no. 2.

Wood, E., M. W. Tyndall, J. S. Montaner, and T. Kerr. 2006. "Summary of Findings from the Evaluation of a Pilot Medically Supervised Safer Injecting Facility." *Canadian Medical Association Journal* 175, no. 11: 1399–1404.

CHAPTER SIX

The Urban Political Pathology of Emerging Infectious Disease in the Age of the Global City

Roger Keil and S. Harris Ali

Straddled across globalized networked webs, cities today are vulnerable to a wide range of global incursions. Among those new threats are emerging infectious diseases (or EIDs), which are, in some cases, like tuberculosis, for example, *re*emerging disorders (Gandy and Zumla 2003). Severe Acute Respiratory Syndrome (SARS) in particular signaled the turn to EIDs as a possibly pervasive and devastating destabilizer of individual cities and the networks in which they are constituted (Ali and Keil 2008). Previously unknown, the SARS coronavirus bridged the gap between animals and humans at some point before November 2002 and worked its way through the wild animal markets of southern China into the hotels and hospitals of some of the world's most hyperconnected global cities, such as Hong Kong, Toronto, and Singapore. When it burned itself out in late summer of 2003, it had killed approximately eight hundred people worldwide and had made many thousands violently sick. Its symptoms are reminiscent of other respiratory illnesses, which originally earned SARS the name "atypical pneumonia" during the early stages of the outbreak. But the course of the disease tended to be more rapid and violent than expected and required immediate hospitalization and machinic support. Its "institutional" character (transmitted in hospitals rather than in the community) turned out to be both an advantage and a disadvantage in devising containment strategies, as the places that were supposed to be the source of help and healing were also the places where the virus spread most efficiently.

In Toronto, where forty-four died and more than two hundred were infected, strategies against the spread of the disease were consequently concentrated on the hospital system. In the meantime, this global city's links to the outside world were fundamentally compromised: conventions and congresses were cancelled, airplanes arrived without passengers, the city's Chinatown suffered devastating losses to its retail and restaurant sector, while the airport was equipped with thermal detectors in what proved to be an ineffective screening strategy to identify carriers of the virus. Most important, the World Health Organization (WHO) at one point issued a travel advisory against Toronto, which called into question business as usual in this most global Canadian metropolis whose economy relies profoundly on the flow of business and tourist travel. The result was a fundamental destabilizing of the city's political institutions as the functioning of multilevel governance in public health was called into question and the political fallout of the epidemic reached into all spheres of life.

We follow the virus through Hong Kong, Singapore, and Toronto—a global city "clique" tied together "as a set of nodes in which each is connected to all the others" (Taylor 2004, 117)—to show both the vulnerability of those cities to emerging disease and the tightly orchestrated measures that were taken at all scales to respond to the previously unknown infectious disease, which killed hundreds and made thousands sick in these three cities. Looking at the practices and policies engaged between local health care providers and the WHO demonstrates that, like one hundred years ago, when the "bacteriological city" (Gandy 2006) created its specific inventions—sewers, fresh water, hygiene programs, public health—to fend off infectious disease, the global cities of today are struggling to create new institutions through which to stabilize the tenuous socionatural relationships on which they are built. These institutional public health reforms and responses are designed to deal with increased and intensified mobilities of various sorts (information, people, microbes) through the rather porous site of the global city. In turn, they emphasize tension between the networked and topological perspectives on cities (chapter 2) and the more conventional conception of cities as closed and bounded entities. SARS in particular has demonstrated the vulnerability of urban networks and individual places within them. At the same time, the urban connectivity that made places vulnerable in ways unknown in previous periods (due to increased mobilities mostly) also created new opportunities to reimagine the urban as a global phenomenon. The case of SARS confirmed the topological construction of the global cities network on one hand (Smith 2003) and the existence of tightly controlled scaled hierarchies of decision making in disease control on the other. In fact, the SARS trajectory can be seen darting across the bumpy conceptual terrain of territory, place, scale, and networks/reticulation (Jessop, Brenner, and

Jones 2008; for an extended discussion of this argument, see Ali and Keil 2006 and Van Wagner 2008). The existing scaled hierarchies of spatialized communities of the health systems of China, Hong Kong, Singapore, and Toronto led to significant problems in communication and information handling during the SARS outbreaks, thus resulting in problematic containment efforts that were somewhat imperfect and slow when dealing with the mechanisms of viral diffusion that simply did not respect the socially constructed and politically guarded boundaries in which the existent scaled public health bureaucracies operated. The SARS outbreaks brought into sharp relief how the loss of clear boundaries was of course a direct outgrowth of global city formation and the growing incapacity of the (local) state in particular to deal with crises visited on an urban region by its growing internationalization. At the same time, it was indeed containment efforts of the conventional sort, based on the view of the bounded city with fixed, spatialized, and restricted access, that proved most successful—namely, contact tracing, isolation, and home quarantine. That is, despite the topological character of disease diffusion in the unbounded city, it was those strategies that were largely, but not wholly, based on the view of the bounded city that seemed to be most effective. Having said this, it was not such strategies alone that led to the success of the containment.

The modern Western city is largely a product of specific—Kaika and Swyngedouw (2000) would say "fetishistic"—territorializations, which hedge in the risks of the bacteriological city in ways that make life in large conurbations possible. Material and social engineering created biophysical, technical, and sociospatial bulwarks against the threat of infectious disease, which used to ravage pre- and early industrial cities (and are still rampant in the developing world). At the same time, urban measures against disease were part of a larger, imperial biopolitical arrangement: "There is no question that public health and hygiene were crucial measures of population and civic management, forming, at every level, imagined and corporeal national communities, as well as communities of colonial extension" (Bashford 2006, 68). The city must be understood as a complex product of reactions to a perilous nature and—we might add—perilous social. The political program of the bacteriological city is not just directed against the germs, which it sought to eradicate, but also against their carriers, which it sought to control. The bacteriological city of the twentieth century was based on an entirely human-centered and purified science, othering animals, disease, and less-than-human humans through its technical ingenuity and government "biopolitics." The city was recast from being the *source* of disease to its antidote, a system set up to fend off disease that was, from now on, associated with backwardness and rural life. Only recently—with SARS and other emerging infectious diseases—has there been a certain reversal of the twentieth century, as infectious disease has returned to the modern city. While purification and biopolitics were the characteristics associated with the hygienic city of the last century, we have now entered a phase in which the reemergence of infectious disease at a mass scale forces us to rethink the relationships among our built environments, our institutional arrangements, and our practices as urban dwellers (Keil and Ali 2007). This includes, as Doreen Massey among others has reminded us, a more-than-human and more-than-urban domain (Massey 2005). This addition is crucial. That we have been punished for the anthropocentric illusion of the bacteriological era has been clear, if it ever really was hidden, at least since the SARS epidemic of 2003 when the other-than-human entanglements of our urban existence became painfully visible (Braun 2008; Jackson 2008).

How does the plague make the city? In two ways: (1) through the consequences of squalor, which is a localized, place effect of disease; and (2) through the consequences of mobilities, which are globalized, at least potentially (Diamond 1999). In a concrete way, SARS was a product of the sharply increased connectivity between places and the accelerated and condensed modes of human interaction found in today's globalized world. In the words of a leading laboratory scientist who played a major role in the discovery of the SARS virus in 2003,

> SARS initially came from markets in Guangdong. Now, we could argue why hasn't this happened before, you know, going back two hundred years. I'm sure it happened, it probably happened many times but what would have happened, you know, it would have affected a few people in a village, a couple of people would have died and it would have burned itself out then, alright. But now, [we] have these animals brought together in these big markets because there's a huge demand for exotic foods. So, it's not one type of animal maybe in a small village market. Now, we are having hundreds of them.
>
> The size of the markets has increased; the interaction of the human population has increased. Okay, so the opportunities for . . . first, the amplification in the markets is one factor . . . Now influenza virus gets into these light poultry markets and it never leaves it. It just continues because it's an ideal situation, right, because you have new poultry coming in and poultry being taken out. So there's always poultry waiting to be infected. So these markets are actually a source of amplification of the pathogen . . . So, this is why, this is why SARS even 50 years ago, of course, wouldn't have happened. It would have started and died. (interview with microbiologist, Hong Kong University, December 2005)

Urban regions are now often characterized as "unbound" (Amin 2004). Subscribing to a relational politics of place, Amin argues that place and politics are "neither a-spatial . . . nor territorial . . . but topological (i.e., where the local brings together different scales of practices/social action)" (2004, 38). This imposition is not without its own problems, as it tends to ignore the steadfast presence of bounded territorialities and associated regulatory institutions. Yet it is nevertheless important in the context of our proposition here as the urbanity exposed by the global SARS crisis spanned different continents, city-states, and communities with their widely variegated political systems, social and cultural habits, and social ecologies. Localism did not apply, as all action was constantly patrolled, controlled, and sanctioned by actors in other jurisdictions, cities in far-flung continents, and organizations of little prior import to the ways of urban living, such as the WHO. In this post-Westphalian age (Fidler 2004), cities were catapulted into the frontlines of global struggles to fend off a new and unknown infectious disease.[1] The political pathology (Fidler 2004) of SARS revealed "a politics of place that is consistent with a spatial ontology of cities and regions seen as sites of heterogeneity juxtaposed within close spatial proximity, and as sites of multiple geographies of affiliation, linkage and flow" (Amin 2004, 38).

Globalization has mostly been viewed as a social and economic phenomenon, widely supported by the most powerful corporations and governments of the world. Global city theory, to a degree, has broken this process down to urban-based actors, politics, and struggles (Keil 1998; Brenner and Keil 2006). When pointing toward global city theory, we do not intend to essentialize a hierarchy or network of particular places that can be traced through tracking indicators and rankings. By contrast, we refer to a broad and growing body of often contradictory literatures of urban globalization (for an overview, see Brenner and Keil 2006), which at one end encapsulates all—"ordinary"—cities (Robinson 2006) and at the other end makes all Global Cities into global cities in a state of generalized globalized urbanization, that is, a perspective that views urban places generally as globalizing cities or cities in globalization, instead of focusing unduly on a select group of cities in an elusive and elite group of Global Cities. We also acknowledge specifically the many critical discussions of the concept (see, for example, McCann 2002, 2004) and the global bias in world cities research that seems to move forward without taking into account local political and cultural dynamics (Keil 1998).

Post-Westphalian Political Ecology

The conditions of reemerging infectious disease in global cities are ubiquitous and specific. Even medical knowledge, the antidote to disease, is both global yet performed in situ. There are scaled "moorings" alongside the "mobilities" that characterize the global urban age (Sheller 2004; Sheller and Urry 2006; Urry 2007). At one level, we may understand the problems that face us in terms of the vulnerabilities exposed by infectious disease as caused by the new mobilities and the solution to these problems as provided through old and new moorings. For SARS, the iconic problem was airline connectivity (see chapter 7); the iconic solution was the hospital and the home quarantine. But it is also possible to argue the opposite: the hotel, the hospital, the high-rise tower were specifically problematic moorings where the virus could be spread, while the usually scaled but now accelerated and networked knowledge production chain of index laboratories and super-mobile scientists saved the day under the supervision of the increasingly self-confident WHO. The context specificity of disease is central to the ways in which mobilities and moorings in the urban world relate to each other in how disease is spread. Says one senior WHO scientist,

> Clearly places like Hong Kong, big airline hubs, when something happens there and you think it might go to the traveling population, then that is all that matters. Clearly that makes you think this could spread, and of course with SARS it was exactly like that. It didn't spread very rapidly from Guangdong because Guangdong doesn't . . . I mean, it has an international airport but it's not that big an international airport and for China, the big ones are in Beijing, Shenzhen, and Shanghai and so . . . So it spread locally first of all, but once it got over into Hong Kong, particularly into a traveling population, then it suddenly went to three or four places very rapidly. And that was the risk factor then, that's what made it an international problem. (interview with WHO medical officer, Communicable Disease Surveillance and Response, Geneva, September 2005)

This tension of mobility and moorings, of movement and quarantine, has been inscribed in the transition from international to global developments since the middle of the twentieth century at least. This world-creation has produced a constant tension between the nation-state space and an imagined and increasingly real "world space" (Bashford 2006). Similarly, policymakers in the vulnerable city are in a bind: they encounter systemic obstacles to innovation as the moorings of previous eras of public health wisdom and the container hierarchy of state-centered command chains vitiates against more mobile and flexible modes of fighting disease.

In Toronto, local politicians scrambled to adjust to new health regulations and a travel advisory handed down to the city by the WHO. They jumped jurisdictional scales in trying to get to what they perceived to be the core of their problematic regulatory situation: the leadership of the WHO itself. The WHO on the other hand walked a geopolitical tightrope as it performed various high danger acts at once: its representatives stared down one of the most powerful national governments in the world as they forced China to open its borders to their scrutiny; they organized scientists and labs in a flat chain of information acquisition and sharing; and they revised the International Health Regulations in real time during the crisis. In this situation, the WHO carefully treaded a minefield of international relations as it asserted its own technical role in the achievement of a global policy consensus:

> WHO, as I say, it has its mandate, one because the member states ask it to do that, or tell it to do that, maybe, but also because we, by being a technical organization, have a certain neutral technical authority. We don't represent a country viewpoint, a single viewpoint, we try to represent a global, neutral, viewpoint of what is happening. So countries can trust us, but that has to be earned, that's not something that countries can simply say to you: you have, or you don't have or we can give to you. That has to be something that over the years, as the organization carries out its work, it either creates that kind of trust that it provides neutral authoritative advice that is useful, or it doesn't. (interview with WHO medical officer, Communicable Disease Surveillance and Response, Geneva, September 2005)

The lessons learned from SARS were ambiguous given the changing global political pathology in which the WHO now operates, where nation-states are not the only points of reference anymore. We are interested in the changing role of cities in the network. A senior WHO official and principal author of the International Health Regulations said this:

> Well, have they changed our view of the world? I'm not sure I'd go so far as to say that, but one of the tenets of the new regulations is that it isn't the disease necessarily that defines whether it is an international problem or not. It's the context in which that disease is happening . . . And that's very much what we are trying to get into the regulations. We are not saying you have to notify anymore just three diseases, but think about where it's happening, how it's happening, who it's happening to, and those things tell us whether we need to take any actions on it as well. (interview with WHO medical officer, Communicable Disease Surveillance and Response, Geneva, September 2005)

From the point of view of the constitution of cities in the web of relationships between territoriality and relationality, the reemergence of infectious disease in urban environments leads us to explore the networked forms of political pathology that now determine life in cities.

What we are suggesting here, then, is a new heuristic lens through which we can conceptually understand the global city system as one that is—inter alia—held together by a political pathology. Akin to the notion of urban political ecology—which always contains both the problem and the solution in socionatural processes that constitute it—urban political pathology is a socionatural set of relationships that occurs in the urban setting. It is at once local and global, social and natural, hierarchical and networked, scaled and topological. When we borrow "political" here from political ecology, we deliberately invoke a constructivist, nonessentialist approach to politics, one that is defined through and born in discursive and material struggles over the meaning of social and socionatural relations. Politics, then, can be both a strategic intervention of an institutionalized network of power (as, for example, through the global public health and local hospital networks; or through the class-biased health care system); and a tactical action on the side of individual and collective actors (whereby some may argue, as Latour [2004] does, that such actor-networks contain the other-than-human objects not usually included in the "parliament" or polity of conventional politics).

Securing the Network: Relationality and Territory after SARS

Understood as a primarily urban disease, SARS connected urban centers with one another through the scaled hierarchies of nation-state territories *and* through the networked topologies, the relational connectivities of human or animal bodies or food moving through the global system. The urban thrust of pandemic disease is a generally visible phenomenon of globalizing societies. Says one WHO infectious disease specialist,

> But that's a general concern now, for plague, yellow fever or even meningitis or measles. We're more and more fearful of urban outbreaks. . . . and that's a great concern of us because it is much more difficult to control outbreaks in urban cities. . . . The population is more and more poor, coming from the rural part of the country, and at the same time there is no change . . . it is easier to travel from the countryside to the urban cities. (interview with WHO infectious disease expert, Geneva, September 2005)

Some of this shift has its results directly in the unhygienic conditions of urban life due to structural underdevelopment and colonialism, but some of it also needs to be linked to increased mobility: "Transport communication and people movement is clearly a factor for facilitating the spread of diseases . . . that's clear" (interview with WHO infectious disease expert, Geneva, September 2005). In the first instance, this has the result of making communities around the world less secure:

> Health is a security issue, in emerging infections if they start spreading internationally are a security issue and they can go to any place that has an international airport immediately or they can move within a country through international airports to other places around that country through air travel. (interview with WHO executive director, Communicable Disease, Geneva, September 2005)

This has changed the map of vulnerability in the sense that we are now looking for different kinds of hot spots inside and between territories, as the increased and accelerated relationalities between places lead to new risks to people:

> No country border can prevent the entrance of an infectious disease. And there's no legislation that can do that either, it's a matter of having good surveillance and response mechanisms where the diseases occur, and where there's a risk that they will enter. (interview with WHO executive director, Communicable Disease, Geneva, September 2005)

During SARS, the territorial logic held up to some degree as the WHO rapidly forged its strategy to combat the newly emerged and unknown disease. Cities were defined by territorial political units in a world dominated, defined, and legitimized by sovereign nation-states. A senior WHO official explains the situation:

> The WHO's entry point into countries is through the federal government and that's where our relationships lie. What governs our international response to infectious diseases is the international health regulations and they were established in 1969, the most recent version of those regulations attempt to minimize the international spread of infectious disease with minimal interruption in travel and trade. And the regulations that are in existence still today are a fixed set of regulations and countries are the only reporting source of information—a federal government or a national government. Information is then taken by WHO and it's published in its weekly epidemiological record and a country then can or cannot request, depending on what they need, support from WHO to deal with the outbreak in their country. These have been revised, and the revision was begun by actually setting up the global outbreak alert and response network with this vision of a world on the alert and ready to respond to infectious diseases. And so, we

actually were in the process of revising the international health regulations from the bottom up through this system when SARS occurred. So
we just rolled out this system and afterwards the regulations then went
on to be finalized in the revised international health regulations. (interview with WHO executive director, Communicable Disease, Geneva,
September 2005)

On the input side, where information is collected, the new regulations
soften the territorial borders both toward other (higher- and lower-scale)
jurisdictions and toward nonstate informants, which is an important
aspect of the post-Westphalian political pathology (Fidler 2004):

> Now what is different in the revised health regulations is that not only
> are countries reporting infectious disease, WHO takes reports from
> anywhere it gets infectious disease reports, including GPHIN [Global
> Public Health Information Network], then verifies that, and then
> there's a step that occurs which is looking to see if it's of international
> importance. . . . If it is internationally important WHO has a role to
> speak with countries to make sure that countries that have had the
> disease occurring understand that . . . it is occurring, it is a threat inter-
> nationally, and we work together using the evidence we have to make
> evidence based recommendations on how to stop the spread. We first
> deal with the country where the disease is occurring, and then we deal
> with others outside those countries. So it's a whole new way of working
> with information coming in from different sources, not just countries,
> and being acted on proactively rather than just being published in a
> weekly epidemiological record. (interview with WHO executive direc-
> tor, Communicable Disease, Geneva, September 2005)

The decision tree and international modus operandi still make cities
dependent on the territorial logic of international diplomatic convention.
The WHO enters cities through countries. Both are treated as contain-
ers that fit logically and hierarchically into each other: "WHO has an
office in almost every country and we work through that office. So that
office facilitates the entry, that office coordinates the activities in the country. . . . We go through China to HK . . . It's a requirement" (Interview
with WHO executive director, Communicable Disease, Geneva, September 2005). Still, while the WHO mostly adhered to the territorial logic
that constitutes the organization, it was also under pressure to react to
non-(nation)-state actors that appeared on its doorstep with demands for
recognition. Perhaps the most interesting case of such deterritorialization
was the attempt by local and regional Toronto and Ontario politicians to
persuade the WHO to lift its travel advisory against the Canadian global

city. A delegation of government representatives, politicians, physicians, and others traveled to Geneva on April 29, 2003, to meet with the WHO director general, Gro Harlem Brundtland. This was considered a highly "unusual" and "nonregular" step by the WHO, as its "relationship is usually with federal governments" (interview with WHO executive director, Communicable Disease, Geneva, September 2005).

From the point of view of the WHO, the most powerful network actor, it was important to note the interplay of central or federal control over surveillance and response mechanisms and the local ways to react in that framework in China and Canada, something altogether different in city-state jurisdictions like Hong Kong and Singapore:

> The most important thing for infectious disease is good surveillance and appropriate response, whether it occurs internationally, nationally, or sub nationally . . . The Canadian governments, like the Chinese governments, had a very difficult problem because their health activities were decentralized, and this is normal in that decentralization is the way health is best conducted, in my opinion. However, the countries with decent systems had a greater difficulty in getting collaboration with health departments in the provinces, in further down, until they made it a civil and non-health issue. And the minute that both China and Canada made it a non-health issue but a central issue with the Prime Minister fully involved, things changed dramatically . . . which just shows that it's important in decentralized systems that this be treated not as just a health issue but as a public security issue. In Hong Kong and Singapore, the reaction was entirely different. Small well— very greatly controlled city states if you will. (interview with WHO executive director, Communicable Disease, Geneva, September 2005)

It was not just spatial considerations that changed during SARS but also the time frame in which the reaction to the new disease was orchestrated:

> It would be fair to say that this is the first global outbreak where there was a 24 hour availability of information and information was continuously coming in through networks of doctors, of clinicians, of virologists, of epidemiologists. So this was evidence based information available in real time to make the recommendations . . . instead of prefixed recommendations they were based on the adaptation through the epidemic. (interview with WHO executive director, Communicable Disease, Geneva, September 2005)

While the WHO has no plans, despite the urban and networked character of SARS, to change its mode of operation by setting up, for example, offices in cities rather than nation-states, their normally hierarchized operations started to develop into a more networked approach that took in information from lower-level governments. In the Ontario case, for example, the provincial and municipal public health offices reported directly to the WHO, which had the effect that early on during the outbreak, the WHO in Geneva was better informed about the disease's presence in Toronto than the federal government of Canada in Ottawa (interview with WHO executive director, Communicable Disease, Geneva, September 2005). The WHO has since taken deliberate steps to increase its understanding of the particular local governance aspects of future pandemic preparedness and response and has sought municipal input into the implementation of its revised International Health Regulations, which were introduced in 2007 (Lyon Biopole and WHO 2008).

The previous discussion highlights the ways through which the international public health organization of the WHO had adapted to the new political pathology brought on by the exigencies of increased global mobility of people and microbes and the "unbounded" qualities of cities and nation-state in the contemporary era. As will be now discussed, the new urban political pathology associated with infectious disease in the global age also led to corresponding adaptations at the urban level, thus reflecting the changes on the other side of the global–urban dialectic.

Global Cities and the Urban Political Pathology of SARS

A defining feature of a global city is its ability to serve as a center for command and control in the world economy based on its position in the network of global cities (Sassen 2000). In turn, the degree of connection and positionality of the global city in the network is based on the inward and outward flows of various types—currencies, people, commodities, information, and so on—that serve to connect the cities. With neoliberalism, the intensification of such flows is seen as desirable and justifiable to integrate the world economy. Following this rationale, political obstacles to the unimpeded movements of flows are removed—and of course, along with that, the possibility for governments to regulate such flows may also be removed as critics of the economic globalization movement contend. However, as was dramatically illustrated in the post-9/11 era with respect to the movement of global terrorists, and in the post-SARS era with respect to the movement of infectious disease, deregulation and reregulation, as well as the encouragement of the "unbounded" city, comes with a price. In response to extreme situations, the potential arises for the development of a crisis politics in which conventional methods of dealing with matters are suspended to adopt extreme measures specifically geared toward the enhancement of "security" and "surveillance," while the social costs of adapting such, sometimes, draconian measures are put aside (see, for example, Wekerle and Jackson 2005; Ali et al. 2006; Ali and Hooker 2006). Thus, another dimension of the political pathology of emerging infectious disease involves dealing with the problematic introduced by the dialectic of fixity with mobility—that is, the fixed stable network of global cities based on firmly established infrastructures for communication, transport, and public health, with the city as a permeable site that is far from fixed because of the constant movement in and out of it (i.e., the city as constantly recreating itself; see Smith 2003). The political problematic that stems from this dialectic is how to regulate the mobility of individuals—and therefore microbes—within the city (chapter 5). Similarly, the correspondingly equivalent problematic for the WHO was the regulation of mobility *between* cities. In this context, crises represent windows of opportunity for governments to (re)gain some control in the regulation of flows, whether in the form of the reinsertion of the actions of a de facto global transnational organization, such as the WHO, or of urban and national governments. The way in which government regulatory actions were exercised in different global cities during SARS were contingent on the contextual politics of place playing out at the different sites. Notably, such a contextual politics of place was influenced by the intersection of the uniquely defined historical developments associated with the particular city in conjunction with the contemporary constraints and enabling opportunities of the urban–global dialectic. Let us consider how the intersection of such forces can account for the similarities and differences in the response and reactions to SARS in the global cities of Toronto, Hong Kong, and Singapore.

Since the positionality of a global city in the network of global cities is one way it may exert influence over the world economy, if a greater number and variety of flows are channeled through a particular global city, it will, in theory, be in a position to exercise greater regulatory power with respect to global economic flows because of its enhanced gatekeeper capabilities. Thus, local governments of global cities often take political and economic measures to ensure that they are well integrated or connected into the various global circuits of capital, information, culture, and so on (chapters 1, 3, and 7). Some global cities, such as London, Tokyo, and New York, are extremely well integrated into the global economy through both inward and outward flows and are referred to as "Hyper Global Cities" (Olds and Yeung 2004). Others, such as Hong Kong and Singapore, are similarly well integrated but are distinguished by having unique capabilities due to their status as "Global City-States" in which the national and urban and local scales are merged together to form a singular and focused UrbaNational development strategy. Notably, the Global City-State has unusually strong capabilities in terms of exerting influence and control within the political economic domain: "Global City-States have the political capacity and legitimacy to mobilise strategic resources to achieve (national) objectives that are otherwise unimaginable in non-city-state global cities. This is because they are city-*states*; they are represented and governed by the state in all of its roles" (Olds and Yeung 2004, 508). By contrast, "Emerging Global Cities," such as Toronto, are somewhat constrained in exercising political economic power because their connection to the global economy is somewhat tenuous, with a far greater reliance on inward flows from the global economy rather than out-ward linkages (Olds and Yeung 2004). Notably, Emerging Global Cities depend on the endowments of institutional resources from higher levels of government, particularly from the national level, which wants to ensure that "their" Emerging Global City is able to play a critical role within the national and regional economies, especially in ensuring that key munici-pal actors and institutions are engaged with the flows required by global capitalism (Olds and Yeung 2004, 506).

The Singapore government's public health response to SARS was praised by the international public health community for its efficiency and effective-ness (Gostin et al. 2003). This success was in part attributable to the unique trajectory of its political history that informed its development as a Global City-State. Following independence from Britain, Singapore then had to contend with political and economic conflicts with both Indonesia and Malaysia, thus leading to an acute public sense of vulnerability (Ow 1984; Lam 2000; Teo et al. 2005). In reaction, the political culture of Singapore has developed into what has been described by observers as centralized, authoritarian, communitarian, and statist (Le Poer 1991; Chua 1995). The influence of such a top-down orientation is clearly seen in the manner in which the Singapore government was able to quickly mobilize political, medical, economic, and public relations strategies to deal with the emerging SARS outbreak (Curley and Thomas 2004; James et al. 2006). For exam-ple, an interministerial committee was quickly set up to decide policies that cut across individual departments, and efforts were coordinated through cross-sectoral interministerial collaboration, which facilitated the efficient communication of information (Devadoss et al. 2005; Ooi et al. 2005). A SARS-dedicated hospital was quickly established while all health care pro-fessionals were required to inform the government of their whereabouts at all times (interview with physician, Travellers' Health and Vaccination

Centre, Tan Tock Seng Hospital, Singapore, January 2006). Further, the government hired a private security firm to enforce surveillance and home quarantine measures of suspected SARS cases through daily "check-up" calls in which those quarantined had to appear before a camera set up in their homes when receiving the call and through electronic wristbands that alerted the agency if home quarantine was broken (Curley and Thomas 2004). Other government action involved the establishment of a dedicated "SARS Channel" on television as part of a campaign to educate and raise awareness about the disease (Ching 2004). The actions taken by the Singaporean government to centralize information exchange, limit the mobility of individuals, raise awareness, and so on, all ultimately led to a high quarantine compliance rate.

In contrast, the ability to limit the mobility of citizens and effectively share and communicate information in the wake of SARS in Hong Kong was quite different from Singapore's experience, even though both share the status of global city-state. The 1997 handover of Hong Kong from Britain to China—in which Hong Kong became a Special Administrative Region of China—had important implications for the city-state's problem-ridden SARS response (Ng 2008). In particular, the lack of integration of formal political and institutional arrangements between China and Hong Kong led to formidable difficulties in sharing epidemiological information and communication.

During the early stages of the outbreak (at which time the causative agent was not identified and the disease was referred to as "atypical pneumonia"), Hong Kong awaited public health reports to emanate from the Chinese government about the situation in Guangdong Province, but none appeared. Moreover, because the Chinese government still tightly controlled the official media in China, the government could not be relied on for valid information. In this connection, a member of the Hong Kong SARS Expert Committee noted that

> at the beginning, there was inadequate communication, when we heard about the rumors of this atypical pneumonia in Guangzhou, when people rushed to buy this white vinegar and then rushed to buy Chinese medicine. We were querying why they rushed to buy this. Because we don't have the direct report going to Hong Kong. (interview with physician, Centre for Health Education and Health Promotion, Chinese University of Hong Kong, December 2005)

The lack of information at the hands of Hong Kong health care professionals made it difficult to combat the disease. In protest, exasperated doctors and nurses went public with their frustrations and called radio stations to criticize the government and the hospital authorities for the dearth of information, as well as to voice their concerns about the insufficient number of masks and other resources required to manage the outbreak effectively. Very little positive information about the control of the illness or the outcome of medical therapy was available to the Hong Kong public through government channels (Lee et al. 2005). Nor did the government direct the people on how to protect themselves, use and dispose of masks, and so on (interview with sociologist, Department of Sociology, Chinese University of Hong Kong, January 2006). In fact, an initial denial of an outbreak by the government followed by the massive implementation of preventative measures may have been construed as a politically driven response, and such an inchoate response would only accentuate the lack of public trust and paranoia about the unknown disease (Lee et al. 2005).

As a result of the political circumstances of the newly reestablished relation between China and Hong Kong, confusion reigned during the early stages of the outbreak, with little trust in official decrees leading to an initially uncoordinated response that was quite different from the experience of Singapore, but much closer to that which unfolded in Toronto.

An Emerging Global City's reliance on higher levels of government (i.e., the national and provincial in the case of Toronto) may either enable or constrain its public health efforts. Constraints may be more likely to be expected if the higher levels of government subscribe to neoliberal directives aimed at the downsizing (and privatization) of collective services while downloading the remaining services to the municipal level without adequate resource support. Such was indeed the case in terms of the Toronto SARS response, as all government-sponsored investigative commissions (at the national, provincial, and municipal levels) noted similar problems that could all be ultimately traced to a lack of resource support due to national government cuts (see also Affonso et al. 2004; Sanford and Ali 2005; Salehi and Ali 2006; Keil and Ali 2007). For example, it was noted by the provincial commission that "SARS showed Ontario's central public health system to be unprepared, fragmented, poorly led, uncoordinated, inadequately resourced, professionally impoverished, and generally incapable of discharging its mandate" (Campbell 2004, 12). The federally commissioned Naylor report identified similar shortcomings, including

> lack of surge capacity in the clinical and public health systems; difficulties with timely access to laboratory testing and results; absence of protocols for data or information sharing among levels of government; uncertainties about data ownership; inadequate capacity for epidemiologic investigation of the outbreak; lack of coordinated business

processes across institutions and jurisdictions for outbreak management and emergency response; inadequacies in institutional outbreak
management protocols, infection control, and infectious disease surveillance; and weak links between public health and the personal health
services system, including primary care, institutions, and home care.
(National Advisory Committee on SARS and Public Health 2003, 1)

Meanwhile, the Medical Officer of Health in Toronto at the time of the
SARS crisis commented in an interview that one of the most significant
constraints during the crisis was the use of an outdated information system built in the 1980s that made case management and tracking extremely
difficult:

> The information systems that we had to use, and then set aside to
> conduct disease surveillance, case management contact tracing and
> management and then quarantine . . . ought to have been our primary
> mechanism for just surely counting the number of cases we had on a
> day-by-day, hour by hour basis and [then] reporting . . . to the Ministry
> of Health, and through them to the national and international public
> health bodies . . . [but was] totally inadequate to do this kind of func-
> tion on the scale that was required . . . There was no money, there was
> no will, there was no time, there was no nothing to replace it, 'cause it
> wasn't a priority until it all fell apart. (interview with Toronto Medical
> Officer of Health, April 2006)

Further information-handling difficulties during the outbreak were identified in the provincial Campbell report (2004, 12) in that "there was a
perception among many who fought SARS that the flow of vital information to those who urgently needed it was being blocked or delayed for no
good reason." Similarly, the federal Naylor report noted that

> the lack of clarity around the flow of communication and the reporting
> structure, the absence of a pre-existing epidemiological unit coordi-
> nated with the local health units and the absence of clear public health
> leadership above the Epi[demiological] Unit provided an environment
> in which the crucial elements of the fight against SARS were discon-
> nected from each other. (National Advisory Committee on SARS and
> Public Health 2003, 1)

The lack of coordination and information handling in the Hong Kong
and Toronto cases, and the success in these respects in the Singapore
case, reveal the extent to which the monitoring and regulation of mobility during an outbreak is a function of context specificity both in terms
of its nature as a global city and contingent political circumstances. By

extension, this observation on failure and success of particular instruments, processes, and measures allows us to recast some current questions of urban and regional governance more generally (Keil and Ali 2007).

Conclusions

Global cities are connected to one another through various flows. This may, on the one hand, be represented as a static, hierarchical (or networked) structure with global cities serving as somewhat fixed loci. On the other hand, the constant movement of flows (including people and microbes) between and within global cities highlights the importance of the particularly dynamic qualities of the global cities network (as well as the global cities themselves). This dialectic between the hierarchical and networked-topological dimensions of the contemporary global city has induced certain changes and introduced new institutional constellations of governance to deal squarely with the emergence of a new disease ecology. That is, one that is informed by new non-nation-state relations and the increased mobilities of the various flows (Ali and Keil 2007; chapter 5). One notable set of changes in this vein are those associated with the adoption of a post-Westphalian logic in which the once unquestioned premise of giving exclusive priority to the unfettered sovereignty of nation-states and the protection of trade becomes less dominant than it once was. This reconsideration in the working logic of post-Westphalian public health was thought necessary by the WHO as well as local public health agencies in light of the contemporary circumstances, namely, the urgent need to respond as quickly as possible to contain the threat of infectious disease in this age of intensified global mobility and interconnectedness. As we have reviewed, the changes in strategic thinking and modus operandi involved in the public health management of infectious diseases were well illustrated by the responses to SARS. Changes could be discerned, for example, in terms of the governance of infectious disease at different scales. Thus, the investigative reports produced by various government commissions often concluded with practical suggestions concerning the "lessons learned" from SARS and how to go about implementing these changes, especially in light of the possibility of a flu epidemic. Changes were also suggested, for instance, in terms of the need to train more infectious disease specialists and to change hospital emergency protocols in light of the recognition that many staff members were no longer available because of self-imposed quarantine under conditions of nosocomial (i.e., within the hospital/health care setting) transmission. At another scale, changes in this post-SARS and post-9/11 era may be seen in the pooling of security/intelligence with public health information and strategies, as the U.S. Central Intelligence Agency and the Centers for Disease Control and Prevention in Atlanta incorporated the lessons from SARS into the continental security hegemony of the United States in North America (CIA 2003; Institute of Medicine 2004). Meanwhile, at the global scale, the post-Westphalian actions of the WHO revealed the susceptibility of local governance systems to supranational intervention. There was never any mechanical logic at work, though, but the dialectics of territorial and deterritorialized dynamics in the responses to the pandemic at various places generated idiosyncratic and perhaps place-specific solutions that pointed toward variegated strategies for future disease threats in the network. In Toronto, for example, the public debate on infectious disease in and after SARS created awareness of the city's vulnerability beyond the immediate crisis in 2003, leading to increased media attention on every subsequent disease threat, such as bird flu or Norwalk virus. This heightened attention parallels a more profound reorientation of health care practitioners in Toronto, a veritable rescaling of responsibilities and the opening up of collaborative possibilities beyond the previously given institutional framework based on the Westphalian order:

> The University of Toronto academic hospitals have gotten together to plan together for pandemic flu, together as a group . . . We never, ever would have done that in the past and we are largely doing this because our experiences during SARS . . . we were all doing different things, our staff were totally freaked out and we're also doing it because we realized that self-reliance is a good thing and if we actually had a bad pandemic, we could not expect the ministry nor Toronto public health to sort of come to our rescue. And so there's a very concrete example of where we actually are . . . taking it much more local than we would expect. Nobody wants to get directives from the ministry anymore without having a lot of say in terms of what those directives are going to say. We all . . . post-SARS . . . I got a lot more involved in work the ministry was doing because I never wanted to be on the receiving end of that again, without having a say in what was going on and the hospital very much supports that. So I'd say absolutely there's been a lot more local . . . local way of thinking about these things. (interview with physician, University Health Network, Toronto, November 2005)

The success of the SARS outbreak response also illuminated the workings of a new urban health "subpolitics," especially as it related to international aspects of disease spread. That is, the SARS crisis exposed a layer of public health governance that resided in the subpolitical regions of health research. This is characterized by the ability of scientists around the world to unite their research endeavors in a virtual laboratory through sharing epidemiological information in real time through modern digitalized information and communication technologies, thereby enabling the global medical community to track, isolate, and characterize the genetic code of the SARS coronavirus quicker than any comparable virus in human history (Ali and Keil 2007).

We return, then, to the notion of an urban political pathology. As we have shown throughout this chapter, flows other than money, workers, and culture begin to define the openness and boundaries of the global city network. In this context, as we have demonstrated here, these other flows, such as pathogens, for example, are embedded in complex hierarchized networks of scaled topologies. They are part of what we recognize now as the urban world in which we live.

Note

1. The reference here is to the Peace of Westphalia in 1648, which ended the Thirty Years War and cemented our modern understanding of national sovereignty as the foundation of international politics. Although the debate on post-Westphalian conditions is rather broad and colorful, when referring to the concept here, we follow Fidler (2004) in denoting a quite specific process by which the international system of nation-state–based regulation has been perforated at both ends by more supralocal institutions. We would add to Fidler's definition that such perforation of the Westphalian logic also occurs from below, that is, from the direction of lower-level states and polities, including the local state.

References

Affonso, D. D., G. J. Andrews, and L. Jeffs. 2004. "The Urban Geography of SARS: Paradoxes and Dilemmas in Toronto's Health Care." *Journal of Advanced Nursing* 45, no. 6: 568–78.

Ali, S. H., and C. Hooker. 2006. "SARS and Security: Public Health in the 'New Normal.'" Paper presented at the 101st annual meeting of the American Sociological Association. Montreal, QC. August 11–14.

Ali, S. H., and R. Keil. 2006. "Global Cities and the Spread of Infectious Disease: The Case of Severe Acute Respiratory Syndrome (SARS) in Toronto, Canada." *Urban Studies* 43, no. 3: 491–509.

———. 2007. "Contagious Cities." *Geography Compass* 1, no. 5: 1207–26.

———, eds. 2008. *Networked Disease: Emerging Infections in the Global City.* Oxford: Wiley-Blackwell.

Ali, S. H., R. Keil, C. Major, and E. van Wagner. 2006. "Pandemics, Place, and Planning: Learning from SARS." *Plan Canada* 46, no. 3: 34–36.

Amin, A. 2004. "Regions Unbound: Towards a New Politics of Place." *Geografiska Annaler* 86B, no. 1: 33–44.

Bashford, A. 2006. "Global Biopolitics and the History of World Health." *History of the Human Sciences* 19, no. 1: 67–88.

Braun, B. 2008. "Thinking the City through SARS: Bodies, Topologies, Politics." In *Networked Disease: Emerging Infections in the Global City*, edited by S. H. Ali and R. Keil, 250–66. Oxford: Wiley-Blackwell.

Brenner, N., and R. Keil. 2006. *The Global Cities Reader*. London: Routledge.

Campbell, A. 2004. *The SARS Commission Interim Report: SARS and Public Health in Ontario*. Toronto: Ontario Ministry of Health and Long-Term Care.

Ching, N. W. 2004. *The Silent War*. Singapore: Tan Tock Seng Hospital Pte Ltd.

Chua, B. H. 1995. *Communitarian Ideology and Democracy in Singapore*. London: Routledge.

CIA [Central Intelligence Agency]. 2003. *SARS: Lessons from the First Epidemic of the 21st Century: A Collaborative Analysis with Outside Experts*. Washington, D.C.: Office of Transnational Issues, CIA.

Curley, M., and N. Thomas. 2004. "Human Security and Public Health in Southeast Asia: The SARS Outbreak." *Australian Journal of International Affairs* 58, no. 1: 17–32.

Devadoss, P. R., S. L. Pan, and S. Singh. 2005. "Managing Knowledge Integration in a National Health-Care Crisis: Lessons Learned from Combating SARS in Singapore." *IEEE: Transactions on Information Technology in Biomedicine* 9, no. 2: 266–75.

Diamond, J. 1999. *Guns, Germs, and Steel: The Fates of Human Societies*. New York: Norton.

Fidler, D. P. 2004. *SARS: Governance and the Globalization of Disease*. Basingstoke, UK: Palgrave Macmillan.

Gandy, M. 2006. "The Bacteriological City and Its Discontents." *Historical Geography* 34:14–25.

Gandy, M., and A. Zumla, eds. 2003. *The Return of the White Plague: Global Poverty and the "New" Tuberculosis*. London: Verso.

Gostin, L. O., R. Bayer, and A. L. Fairchild. 2003. "Ethical and Legal Challenges Posed by Severe Acute Respiratory Syndrome: Implications for the Control of Severe Infectious Disease Threats." *Journal of the American Medical Association* 290, no. 24: 3229–37.

Institute of Medicine. 2004. *Learning from SARS: Preparing for the Next Disease Outbreak*. Workshop summary. Washington, D.C.: National Academies Press.

Jackson, P. 2008. "Fleshy Traffic, Feverish Borders: Blood, Birds, and Civet Cats in Cities Brimming with Intimate Commodities." In *Networked Disease: Emerging Infections in the Global City*, edited by S. H. Ali and R. Keil, 281–96. Oxford: Wiley-Blackwell.

James, L., N. Shindo, J. Cutter, S. Ma, and S. K. Chew. 2006. "Public Health Measures Implemented during the SARS Outbreak in Singapore 2003." *Public Health* 120:20–26.

Jessop, B., N. Brenner, and M. Jones. 2008. "Theorizing Sociospatial Relations." *Environment and Planning D: Society and Space* 26:389–401.

Kaika, M., and E. Swyngedouw. 2000. "Fetishizing the Modern City: The Phantasmagoria of Urban Technological Networks." *International Journal of Urban and Regional Research* 24:122–48.

Keil, R. 1998. *Los Angeles: Globalization, Urbanization, and Social Struggles.* Chichester, UK: Wiley.

Keil, R., and S. H. Ali. 2007. "Governing the Sick City: Urban Governance in the Age of Emerging Infectious Disease." *Antipode* 40, no. 1: 846–71.

Lam, N. M. K. 2000. "Government Intervention in the Economy: A Comparative Analysis of Singapore and Hong Kong." *Public Administration & Development* 20, no. 5: 397–421.

Latour, B. 2004. *Politics of Nature: How to Bring the Sciences into Democracy.* Trans. C. Porter. Cambridge, Mass.: Harvard University Press.

Lee, S., Y. Y. Chan, M. Y. Chau, P. S. Kwok, and A. Kleinman. 2005. "The Experience of SARS-Related Stigma at Amoy Gardens." *Social Science and Medicine* 61:2038–46.

Le Poer, B. L. 1991. "Singapore: A Country Study." In *Area Handbook Series.* 2nd ed. Washington, D.C.: Library of Congress, Federal Research Division. http://countrystudies.us.singapore/ 52.htm.

Lyon Biopole and WHO. 2008. "Concept Note: International Technical Conference: Cities and Public Health Crises." Lyon, France: WHO Office.

Massey, D. 2005. *For Space.* London: Sage Publications.

McCann, E. J. 2002. "The Urban as an Object of Study in Global Cities Literatures: Representational Practices and Conceptions of Place and Scale." In *Geographies of Power: Placing Scale,* edited by A. Herod and M. Wright, 61–84. Cambridge, Mass.: Blackwell.

———. 2004. "Urban Political Economy beyond the 'Global' City." *Urban Studies* 41, no. 12: 2315–33.

National Advisory Committee on SARS and Public Health. 2003. *Learning from SARS: Renewal of Public Health in Canada.* Ottawa, ON: Health Canada.

Ng, M. K. 2008. "Globalisation of SARS and Health Governance in Hong Kong under 'One Country, Two Systems.'" In *Networked Disease: Emerging Infections in the Global City,* edited by S. H. Ali and R. Keil, 70–85. Oxford: Wiley-Blackwell.

Olds, K., and H. W. Yeung. 2004. "Pathways to Global City Formation: A View from the Developmental City-State of Singapore." *Review of International Political Economy* 11, no. 3: 489–521.

Ooi, P. L., S. Lim, and S. K. Chew. 2005. "Use of Quarantine in the Control of SARS in Singapore." *American Journal of Infection Control* 33, no. 5: 252–57.

Ow, C. H. 1984. "Singapore: Past, Present, and Future." In *Singapore: Twenty-Five Years of Development*, edited by P. S. You and C. Y. Lim, 366–85. Singapore: Nan Yang Xing Zhou Lianhe Zaobao.

Robinson, J. 2006. *Ordinary Cities*. London: Routledge.

Salehi, R., and S. H. Ali. 2006. "The Social and Political Context of Disease Outbreaks: The Case of SARS in Toronto." *Canadian Public Policy* 32, no. 4: 373–85.

Sanford, S., and S. H. Ali. 2005. "The New Public Health Hegemony: Response to Severe Acute Respiratory Syndrome (SARS) in Toronto." *Social Theory and Health* 3:105–25.

Sassen, S. 2000. *Cities in a World Economy*. 2nd ed. Thousand Oaks, Calif.: Pine Forge Press.

Sheller, M. 2004. "Mobile Publics: Beyond the Network Perspective." *Environment and Planning D: Society and Space* 22:39–52.

Sheller, M., and J. Urry. 2006. "The New Mobilities Paradigm." *Environment and Planning A* 38:207–26.

Smith, R. G. 2003. "World City Topologies." *Progress in Human Geography* 27:561–82.

Taylor, P. J. 2004. *World City Network: A Global Urban Analysis*. London: Routledge.

Teo, P., B. S. A. Yeoh, and S. N. Ong. 2005. "SARS in Singapore: Surveillance Strategies in a Globalising City." *Health Policy* 72:279–91.

Urry, J. 2007. *Mobilities*. Cambridge: Polity.

Van Wagner, E. 2008. "Toward a Dialectical Understanding of Networked Disease in the Global City: Vulnerability, Connectivity, Topologies." In *Networked Disease: Emerging Infections in the Global City*, edited by S. H. Ali and R. Keil, 13–26. Oxford: Wiley-Blackwell.

Wekerle, G. R., and P. S. B. Jackson. 2005. "Urbanizing the Security Agenda: Anti-Terrorism, Urban Sprawl, and Social Movements." *City* 9, no. 1: 33–39.

CHAPTER SEVEN

Airports, Territoriality, and Urban Governance

Donald McNeill

> Airports reconfigure geography according to the spatio-temporal
> rhythms and cross-modal standards of global capital, by flattening
> all difference into manageable, measureable and commodifiable
> contours. Airports are geo-mechanical-digital forms that are
> changing the contours of land, sea and sky. To consider the
> relationship between an airport and its environs is to consider the
> entwining of movement, money, land, sky, matter and information.
>
> (Fuller and Harley 2004, 103)

Over the past three decades, there has been a growing interest among governments in the privatization of their airports, thus ending state control
of major airports, reflecting a worldwide shift in government attitudes
toward airports as major drivers of capital accumulation. Evolving from
simple airfields with small terminal buildings in the early days of commercial aviation, the contemporary airport is a highly complex megastructure,
which, in many cases, derives approximately half of its income from its
original functions, such as handling air passenger traffic. From extensive
terminal retailing to hotel and convention centers, along with logistics,
cargo, and aeronautical engineering functions, the understanding of how
the airport platform can be reconfigured is advancing rapidly. Some commentators are now speaking of "airport cities" or the "aerotropolis,"
assigning these spaces with attributes often associated with established
norms of city space (e.g., Kasarda 2004, 2006). These "revenue enve-
lopes" are also managed in ways that differ from state agencies of old.

In part, this is because of the management modes adopted by privatized corporations that own or manage the airports.

An important feature has been the rise of globally operative airport management firms that have long been territorially embedded around the development of major home airports. The Dutch-based Schiphol Group, which now has financial interests in airports worldwide, is partially owned by local state governments in the Amsterdam area. Similarly, the Singapore Civil Aviation Authority, primarily a public body dedicated to managing and developing Changi Airport, has used its proximate relational ties to China by signing management or consultancy contracts in the country's rapidly expanding aviation sector. In each case, airport managers will bring their domestic experience and skills to the management of these foreign airports. Airport management and design is thus an important aspect of the "global intelligence corps," knowledge mediators described by Rimmer (1991) and Olds (2001) and similar to those discussed in chapters 2, 3, and 4: management consultants, real estate advisors, and in-house management groups advise airports worldwide on anything from physical master planning, to logistics design, research, and "strategic visioning." Commissions to design airport terminals are now coveted by elite architectural firms, from Foster and Partners (Beijing) to Renzo Piano Building Workshop (Kansai), and specialist aviation groups from major firms (such as Hellmuth, Obata, and Kassabaum and Skidmore, Owings, and Merrill).

The growing complexity of airport governance comes at a time of more general reflection on the nature of urban territoriality. We have seen ontological debates over the nature of cities, as a "spatial fix," or "fetish," a means of projecting social relations onto a defined territory to make them more understandable (Beauregard 1993; Pile 1999; Amin and Thrift 2002). Cities, and urban, are often used interchangeably, which is problematic when trying to analyze the "spatial formations" that we call cities. Airports may act as a useful heuristic in attempts to work out just how global cities may be interrelated; they may, ultimately, be seen as innovations in themselves, in the construction of a globally synchronized urban time-space. As Fuller and Harley hint in their book *Aviopolis* (2004), airports may simply be *too* urban for the city. In Sydney, they argue, for more than eighty years, the airport has "been 'terraforming' its environs, sucking highways and rail corridors towards it, re-zoning its surrounding suburbs, flattening houses and changing the geography of the city around it" (41). We can begin to get an idea of airports as a scale beyond what we normally associate with city life, and it is clear that few major international airports are controlled by an elected city council.

Given their significance as enablers of territorial mobility—of people, of goods, and of the policy knowledge discussed in the other chapters in this volume—airports deserve a central role within debates about the intense material mobilities that constitute contemporary society, and their impact on existing territorial formations. Airports have always had as their main point of existence the management of relationships between places, usurping rail and seaports as aviation technology allowed the reconfiguration of interplace space-time. Yet what is interesting about airports is how their relational rationale has been accompanied by the development of dense territorial formations of sociality and economic activity. As a workplace, as a surface transport node, as a site of new office developments and business conferencing, the major airport is now a significant site of urban sociality in its own right, where people commute to and from, and linger in, rather than merely passing rapidly through as the "gateway" metaphor may imply (see chapter 6).

The chapter begins by conceptualizing the spatiality of airports, by considering their relationship to territory, and to sociality, particularly as a workplace. It then examines three key strands of airport management: (1) the shifting spatiality of airport management, especially under a regime of privatization; (2) the problematic governance of airports in the context of their city–regional scale of operation; and (3) the relationship of airlines to airports, which emphasizes the complex interplay of territoriality and mobility that air industry managers are required to negotiate. The chapter concludes by considering how airport territory is now subject to a multiple and overlapping set of ownership regimes, systems of operation, and formal territorialized legal frameworks, with important implications for understanding the material impact of airports on the urban fabric, the embodied nature of travel, and the territorial "stickiness" that accompanies these spaces of hypermobility.

Conceptualizing the Territoriality of the Airport

The story of airport development is one of some significance in the development of the contemporary nation-state. Often conceived as an object of great pride by prime ministers, military leaders, or boosterist regional politicians, and underpinned by a popular culture of travel and mobility (Urry 2000; Cresswell 2006), the airport has a powerful symbolism in national territories. While often understood as a facilitator of globalization, it is still often conceived in geopolitical terms both as a means of uniting large, dispersed national territories, as well as engaging with extraterritorial actors. In Malaysia, for example, the new Kuala Lumpur

International Airport was sited in Sepang, at the center of an emerging high-technology corridor that sought—along with a repositioning in the regionalized global economy—to form a "'backbone' controlling the national geo-body" (Bunnell 2004, 109). In the United States, the construction of a federal system of airports was a fundamental part of the country's postwar development and growth. It has also become clear that a small number of city-states have emerged, particularly Singapore, Hong Kong, and Dubai, which position themselves, and their nation-building strategies, in the context of global flows of passengers and goods through open-skies policies.

However, while major international airports are often seen as the responsibility of central government, the urban governance of these airport territories is complex, given their location within uneven, historically evolving structures of state control (Humphries 1999; Caruana and Simmons 2001; Güller and Güller 2003). Embedded within large metropolitan regions, airports are, in some places, integrated within strategic regional planning frameworks but are, in other places, configured as enclaves, beyond the conventional planning regime of local and regional authorities. Furthermore, the varying powers of state planning and development regimes dictate the ability to carry out major redevelopment schemes. In London, the scale of public opposition to the development of Heathrow's Terminal 5 led to the longest running planning inquiry in British history (Griggs and Howarth 2004). Meanwhile, in states with highly centralized powers, new airports, such as Hong Kong International Airport, can be planned methodically and with a minimum of opposition. A brief survey of airport development across the world would show that despite some surface homogeneity, airports are highly complex in their configuration, evolution, and operation. However, it is important to explore these questions through a thoroughgoing analysis of the corporate, financial, and political geographies of airport governance. I suggest that several conceptual issues require some foreshadowing.

First, the airport sits in an interesting place in the intersections between territoriality and relationality. In *Aviopolis* (2004), Fuller and Harley indicate the peculiar logics of airport geographies. They nominate airports as "terraformers," a felicitous term that highlights airports' huge impact on land and environment but that also hints at the way in which relational understandings of territory can be conceptualized. Take, as a point of comparison, rail travel, which—as a network—acts to transform place (Schivelbusch 1986; Bishop 2002) by overcoming the intervening territories that separate the origin and destination of the train. Here, the concept of the corridor is important: "A corridor 'gathers' the elements of the landscape and culture, thereby creating new places, perspectives, meanings and experiences, both around it and, importantly, within it" (Bishop 2002, 299). Drawing on Virilio's (1991) counterposition of the vector—a one-dimensional line of speed and direction—with the stickiness of territory, Bishop argues that "just as the vectorial qualities of the rail corridor (the immensity of its length in contrast with its extreme narrowness, its evocation of mobility, of an efficient journey, its focus on departure and destination) constantly undermine local-territorial or place-related aspects, so too there are moments of intense resistance to, even subversion of, the dominant vectorial definition" (299).

Air travel carries out a similar process, melding landscapes, particularly in nation-states with large internal spaces. In so doing, airports possess an affective quality, which implicates it in the "banal nationalism" of the nation-state (Billig 1995). Airport usage became one of the defining social practices of modernizing urban societies, a mundane social practice, if not a quotidian one. For Adey (2006), this can be captured in the concept of "air-mindedness," which signifies the "organizing complex or ideal that orchestrated the enthusiasm and development for aviation. It distributed politics, places and things for the development of flight and the betterment of the nation" (346). And yet unlike more leisurely forms of rail travel, the focus is on the destination rather than the journey. While the railway journey allows for a "sideways glance, acknowledgement of territory, nostalgia for immobility, depth and place" (312), air travel thus effaces the role of landscape in the formation of geographical imaginations. It is worth reflecting on whether airports have undermined modern ideals of cityness and have contributed to the stretching out of urbanized social relations.

Second, consequently, air travel is often seen to be deterritorializing, unlike other forms of mass travel. This is partly due to the escape of the tyrannies of land travel, where new routes are a matter of negotiating airspaces rather than the expensive, time-consuming, and often controversial process of constructing new motorways or rail links. For international travelers who use airports as a hub to change planes, this becomes very evident: "Many . . . frequent flyers have, of course, never really been to Singapore, LA or Frankfurt. They never leave the airport. Singapore, for instance—whatever entity that might be—remains abstract, a concept gleaned through newspaper reports, stories told by friends and colleagues, and by time spent at its airport. They may never have been 'to Singapore' but they have been 'in Singapore'" (Fuller and Harley 2004, 39). In this state, cities become known through the three-letter code allotted to them by the International Air Transport Association (IATA). Such codes are designed to allow for the easy identification of connected ports and, as such, serve to conceptualize airports as relational nodes.

Third, airports can be seen as "large socio-technical systems" (Graham and Marvin 2001), which, in turn, are often seen as "black boxes." This is particularly so given the levels of security and secrecy that often govern their operation, obfuscation now increasingly due to commercial competition rather than bureaucratic paranoia (see, for example, the collection of papers in Salter 2008; Walters 2002; Adey 2003). However, in major privatized airports, there can be added problems for management, as complex sets of stakeholders with sometimes competing goals (trade unions, airlines, contract caterers, baggage handlers, air traffic controllers, taxi drivers, local residents, politicians, government immigration and customs officials, and government aviation regulators) are involved in their governance. Airports are famously populated by armies of underpaid and "flexible" labor, often living in the neighborhoods that lie beneath the approaches to major airports. The logistics of coordinating this huge and highly differentiated labor market of pilots, air traffic controllers, plane engineers, security staff, baggage handlers, catering staff (both in the terminal and for in-flight food) are incredibly complex, and industrial action by any one of these groups can quickly bring the airport machine to a shuddering halt. However, the complex web of subcontracting and outsourcing that characterizes airports as multiemployer workplaces is often underpinned by a strong degree of trust and cooperation among workers, rather than antagonism (Carroll et al. 2005; Marchington et al. 2005). Thus, despite the temptations involved in seeing airports as abstract, depersonalized spaces, they might best be seen as a metaphor for the significance of embodied social practices to the successful functioning of social space.

The Globalization of Airport Governance

As already noted, the uneven shift toward the privatization of airport management has opened up a range of opportunities for companies interested in infrastructure investment. This has shifted airport management from urban governance (by an elected state), to corporate governance, thus undermining linkages between airport operation and urban territorial goals. As Graham (2003) notes, before privatization, major airports such as Frankfurt (45 percent of which is owned by the state of Hesse, 29 percent by the city of Frankfurt, and 26 percent by the federal government) and Vienna (50 percent owned by the national government, 25 percent by the province of Lower Austria, and 25 percent by Vienna city council) were neatly nested within a traditional scalar hierarchy of governments. There was thus a clearly identifiable link to a territorialized state, with some degree of political accountability. Privatization legislation in different countries replaced such accountability with differing types of regulatory structures, often with limits to the quantity of foreign share ownership, which affect their ability to raise capital and provide pressure on reinvestment of revenues in capital improvements and maintenance, for example (Graham 2003; Doganis 2006).

As a result, the globalization of airport *management* is closely related to the changing nature of the financing of airport *ownership*. The desire of state governments to relieve themselves of the debts and overheads associated with such maintenance has given rise to a globalized infrastructure equity market, contributing to a scenario of "unbundled infrastructures placed in global infrastructure funds that are managed by specialized financial institutions" (Torrance 2008, 2). While these processes are by no means universal, they are of growing significance to urban territorial governance debates and point to the importance of internationally implemented formats of public policy and the in-house risk models and contracts that determine whether a piece of infrastructure be deemed profitable and hence realizable. This further complicates governance agendas and poses conceptual challenges. As such, "the practices of power may be less about the visible machinery of decision-making and rather more to do with the displacement of authority, the renegotiation of inducements, the manipulation of geographical scales and the mobilization of interests to construct politically meaningful spatial imaginaries" (Allen and Cochrane 2007, 1171).

It is useful to briefly profile some of these companies in a little more depth, through the examples of Macquarie Airports Group (MAp), Schiphol Group, and British Airports Authority Ltd. MAp emerged as a subsidiary of the Macquarie Infrastructure Group, which, in turn, was a subsidiary of Macquarie Bank, the Australian investment bank that has become an increasingly vigorous player in international capital markets since the late 1990s. MAp's most significant move was the successful bid for a long-term lease of Sydney Airport under the Australian Airports Act, which privatized most of the country's major airports. Selective expansion abroad included the purchase of shares in Brussels, Rome Leonardo da Vinci, Birmingham, and Bristol airports. The interest in airports was but one of the bank's growing number of investments in myriad leisure, utility, and transport infrastructure sectors, from bowling alleys to toll roads. Intriguingly, management fees for its airports play an important role in feeding the bank's revenues, allowing it to benefit at several stages in the acquisition process (Jefferis and Stillwell 2007; McLean 2007).

By contrast, the Schiphol Group has taken a broader approach toward value adding. The group "views an airport as an AirportCity: a dynamic environment in which people and businesses, logistics and shops, information and entertainment come together and strengthen each other" (http://www.schipholgroup.com). Its strategy document "Long Term Vision for Schiphol Group" centers on the firm's expansion plans around Amsterdam Schiphol:

> We intend to make Amsterdam Airport Schiphol a leading global airport. Our strategy focuses on consolidating and improving the competitive position of the main port Schiphol as an important intercontinental hub that in terms of connections can compete with London, Paris and Frankfurt.
>
> The airlines—and network carrier KLM in particular—that fly from Schiphol have extensive networks that connect the Netherlands and the Randstad conurbation with all major centres in Europe and the rest of the world. These connections ensure that the Netherlands and the Randstad are linked up with all major hubs that are part of the European and global economy. In the globalising economy it is of vital importance for our continued prosperity that the Randstad continues to enjoy a direct connection with the world's major economic regions and that it remains an attractive business location for international and foreign companies. Schiphol is also an important gateway for tourists travelling to Amsterdam, the Netherlands and the rest of Europe. The network of economic activities located at the airport and within its environs constitutes the major source of employment for the northern section of the Randstad area. (http://www.schipholgroup.com)

Schiphol Group has two subsidiaries: Schiphol International BV, and Consumer International. The group has embarked on a strategy of forming alliances with a small number of airports, with the goal of using the experience derived from Schiphol to add value to its overseas investments. This has three aspects: management operations; the operation of branded AirportCity 'products'; and real estate development. Central to this is the group's AirportCity concept, which seeks to integrate aviation and nonaviation activities. It has three overseas management contracts or "participating interests," most notably at Terminal 4 of John F. Kennedy in New York, where a group subsidiary, Schiphol Group USA, has had a 40 percent stake in JFK International Air Terminal LLC since 1997, and has a twenty-five-year management contract on the new terminal that opened in 2001. Significantly, this was the first non-U.S. airport operator to manage a U.S. terminal. Further interests have been developed at

Brisbane International, in Australia (15.6 percent stake), and at Aruba's airport. The group has developed its AirportCity master planning concept in Stockholm Arlanda and Jakarta International, operating joint venture companies focused on expanding retail, hospitality, and accommodation facilities within the airport terminals. It also has significant "logistics park" real estate holdings (as joint ventures) in Milan's new Malpensa Airport, and at Hong Kong International Airport's Tradeport. Furthermore, it is worth noting that the group also owns Dartagnan Biometrics Solutions, which markets biometric identification technology, a leading edge security technology within terminal development.

Such forays are not a fully developed aspect of globalization. While a closely observed aspect of the Macquarie Bank model has been its interest in infrastructural entities that can provide reliable revenue streams, the financial climate of the late 2000s has called such strategies into doubt. Furthermore, private airport operators are still tightly controlled by central governments, both in the formal legislative framework that accompanied the privatization process, and in terms of government oversight committees. A key example is the scrutiny faced by BAA PLC, the owners and operators of a number of major British airports. The highly leveraged takeover of the already privatized firm in 2006 by a private equity group led by the Spanish-owned Ferrovial firm was accompanied by a crescendo of complaints over a perceived decline in service quality. In addition, interest groups from across the spectrum—from locals protesting against a planned new runway, to business lobby groups such as London First warning of the dire effects for the British economy that mismanagement was bringing, to eco-campers protesting against the airline industry's contribution to global warming—resulted in an in-depth examination of the firm by the government transport committee, as well as a damning review by its Competition Commission, which has the power to enforce a breakup of companies seen to possess an unhealthy monopoly. In 2008, the UK government's Competition Commission released an interim review into BAA's ownership of the southeast of England's major airports, and issued a warning that this was a quasi-monopoly. The subsequent decision by BAA to sell London Gatwick (given fears that a protracted legal battle would hinder its attempts to develop the other airports in its portfolio, particularly Heathrow) highlights the significant role that the state plays in the regulation of airport ownership, despite the autonomy that many privatized airport management companies appear to possess.

The Airport and the Politics of the City–Region

So what does this all mean for the governance of airport cities? As I shall describe, airports are the machines by which the packaged and commodified vectoral movements—via the airlines—are enabled for the movement of people and goods. As pinch-points in the national territorial fabric, they allow an enhanced mode of surveillance and monitoring that long, exposed land borders are less efficient in providing (see chapter 6). However, ports—both air and maritime—are "scaling up." While old maritime ports are being reconnected with their immediate hinterlands through waterfront residential redevelopments, heritage sites, cycleways, and footpaths, new container ports are increasingly peripheral to dense urban settlement. Just as older airports such as Chicago's Midway or Sydney's Bankstown were once located away from residential development, they have been swallowed up by urban growth (Bouman 1996; Bruegmann 1996). As a response, most new airports are now being planned in-between major settlements, such as Kansai (between Kobe, Osaka, and Kyoto, in Japan) or Kuala Lumpur International Airport (between Kuala Lumpur and Putrajaya, in Malaysia). They also appear in "the middle of nowhere," such as Denver International Airport, a hub and spoke system that far outweighs Denver's importance as a city, but where small planes feed off large jets to serve midwestern regional destinations, thus providing a pure expression of central place theory within a national territorial space.

This has two major dimensions. First, in a deregulated system, airport expansion can be an integral part of localized interurban competition. For example, Los Angeles International (LAX) is an important anchor for the Californian economy, particularly in terms of manufacturing and services import and export. As Erie (2004) describes, LAX has been an understated but fundamental aspect of the postwar growth of Southern California into one of the world's strongest regional economies:

> In 1995, well over half of regionally produced exports (by value) were shipped by air. Airborne exports added more in value to the local economy than did waterborne exports. High-technology, high value-added manufactured products were especially conducive to air shipment . . . [I]n 1994, LAX claimed to generate $43 billion in regional economic activity—roughly 10 percent of the region's then total output—and nearly four hundred thousand jobs. (Erie 2004, 173)

This requires a more sophisticated understanding of the specific nature and composition of air cargo in the context of the new economy: "The relationship between information technology and transportation systems appears to be one of synergy rather than substitution. Thus, while e-mail is reducing overnight air-express deliveries, e-commerce is generating unprecedented demand for air-cargo deliveries" (34). However, the growing economic significance of air transport for Southern California intensified territorial political conflicts between a coalition of "not in my backyard," or NIMBY, groups who sought a limit on LAX expansion, on the one hand, and the city of Los Angeles and a range of "wannabe" councils who sought expansion of the regional airports at the expense of LAX's capacity, on the other (see Erie 2004, chap. 7, for a full discussion). What is interesting in comparison with the UK case, either directly or through delegated state agencies, however, is that the airports remain largely under municipal control. Yet these localities—as was the case with LAX—have to play competitive games of interurban jockeying to retain their preeminence.

Second, major airports in leading regions may be defined by national governments as "state significant infrastructure." In early 2007, the UK government announced its intention to press ahead with the construction of a third runway and sixth terminal at Heathrow (this would allow the privatized entity, BAA, to submit a development application, though the firm would have to raise the finance at market rates). The government's rationale was based on the argument that the national economy depended on Heathrow's ability to handle large numbers of business travelers and was a response to the apparent diversion of passengers to other European hubs such as Paris and Frankfurt. However, there are many contrary views. One of the key issues is its "noise footprint," which carries a sixty-three-decibel limit over a 19-kilometer by 3.5-kilometer swath of territory around the airport, much of which is dense residential development (Hall and Hall 2006). Successive mayors of London have opposed expansion for this and other reasons. In 2008, the business lobby group London First argued for a cut in capacity, suggesting that this would reduce some of the pressures on air traffic control, plane handling, and passenger processing and would allow the airport to operate more efficiently. Ultimately, many feel that Heathrow should be gradually phased out. The increasingly congested airport is served by only two main runways, whereas most airports of Heathrow's size (in terms of passenger numbers) use at least four runways. One solution would be to build an entirely new airport in the Thames Estuary, along the lines of Hong Kong's Chep Lap Kok, where state authorities modified an entire island to form the basis of a new airport platform (Hall and Hall 2006). However, the UK government continues to mobilize discourses of national interest to underpin its intention to expand the airport's capacity, as shown in its white paper on the future of aviation (Department for Transport 2003), and its subsequent approval of an additional runway for Heathrow, and enhanced capacity for London Stansted.

It should be clear, therefore, that the territorial politics of airport development and management is closely connected to the intensity of interplace relations that a nation-state or region enjoys. Booming city–regions generate a strong demand for airline services; in turn, this generates a partly speculative economy of air service supply that airline and cargo firms seek to profit from. These economies underpin the successful interconnection of airports (and their territories), which is central to debates on relational urbanism.

Airlines and Airports

The development of the global air network has also transformed countries and economies, big and small. Singapore, for instance, has grown into a business hub because of our connectivity. Today, Changi Airport has 71 international airlines which link us to 162 cities in 52 countries. This network has made Singapore the regional headquarters for over 60% of the 7000 MNCs present here. By plugging ourselves directly into the global economy, Singapore overcomes the constraints of its geography. (Singapore Prime Minister Goh Chok Tong 2004)

The significance of airline networks as linking mechanisms for intercity networking has been identified by globalization theorists (Keeling 1995; O'Connor 1995, 2003; Beaverstock et al. 2000; Smith and Timberlake 2001). However, what these "direct flight" connections mean for territorial development requires careful attention. As the Singaporean prime minister tells us, connectivity and being "plugged in" aeronautically are key to escaping geographical limitations (but sometimes with complex consequences, as chapter 6 discusses). The importance of "port-city" hub status is also evident in Dubai, one of the most forward moving of the United Arab Emirates. After launching its airline emirates, ostensibly against an open-sky climate, but with cheap government sovereign fund credit allowing for rapid expansion of routes, staff, and airlines, the Dubai Airport Authority was able to maximize its ties with its national "flag carrier," as well as to leverage its geographical centrality to major regional labor markets, both within the Middle East but also as a switching point between Africa, West Asia, and Central Europe. It is worth noting that sovereign wealth funds—state-owned investment agencies that have become particularly active in contemporary capitalism—have been an important feature of the infrastructure investment landscape in recent years. The point was not lost on Maurice Flanagan, CEO of Emirates Airlines, who was responding to the anti-Arab backlash brought about by the successful purchase of several American commercial ports by DP World (Dubai Ports World) from P and O, in early 2006: "Dubai is 'perfectly located to be the global hub of this century,' said Mr. Flanagan. 'Look at a map of the world, with the Americas down one side and China and Japan down the other,' he said. 'If you balance that all on a point, that point is Dubai'" (Timmons 2006). So airports and airlines are both aware of the importance of geographical location, and the significance of connectivity.

However, despite the rhetoric of globalization that is easily attached to the aviation and airport industries, taken together, the deregulation of airport management companies and the restructuring of the airline industry (with its recent history of takeovers, failures, and mergers between national flag carriers) has left behind a tangled web of national regulatory regimes, transnationally owned airlines, and privatized airports. As Graham (2003, 132) puts it, "the airline-airport relationship is starting to become much more to do with the linking of two privately owned international companies, rather than two state owned organizations operated within the limits of national laws and regulations." In large national territories, where divestments of airport and airline management have been pillars of privatization agendas—especially in the UK, and to a lesser extent in Australia and the United States—this makes the particular assemblage of political and corporate interests highly volatile. It is worth noting, however, that in so-called city–state airports, primarily Singapore and Dubai, the national flag carrier and airport are two of the most important institutions in the territorially small polities, and the airport is seen as an important revenue generator for the government as a whole. As such, landing charges at these airports will tend to be lower, to incentivize tourist stopovers and cargo switching.

These relationships are driven by a complex set of regulatory agreements, including general agreements (and disagreements) over prices such as the 1944 Chicago Convention on pricing, bilateral agreements such as the UK–U.S. Bermuda 2 agreement, and a whole raft of nationally specific—and sometimes airport-specific—privatization regulators, and slot allocation methods, most notably the International Air Transport Association (IATA)'s Schedule Co-ordination Conferences (Graham 2003, chap. 5). Here, flag carriers still hold an advantage in many cases through the "grandfather rights" principle, and new entrants (such as Emirates) may find it difficult to break into existing routes.

Nonetheless, major airlines are also concerned by overdevelopment of nonaviation activities at airports, particularly where it interferes with their ability to process passengers efficiently. Delays caused by inadequate security provision, congested terminals, and poorly managed baggage handling facilities can affect the reputation of the airline rather more than the often background presence of the management company. An example of such concerns relates to BAA's management of London Heathrow. In its submission to the UK's aviation regulator, the Civil Aviation Authority, in 2007, Virgin Atlantic cited the example of a recently rebuilt garage in Terminal 3. "As the customer of a monopolist," they tersely began, "Virgin Atlantic faces daily problems in dealing with BAA at Heathrow and Gatwick . . . Despite extensive experience in the construction of airport car parks, BAA has managed to provide a facility that has resulted in a very low level of customer service. At peak times there is virtual gridlock" (7). This posed two problems for the airline: first, customers get poor value for money, which "potentially puts us at a competitive disadvantage"; second, reworking the garage will be expensive, which in a monopoly situation will be passed to customers, they alleged (8). Virgin went further, complaining that the Civil Aviation Authority was unwilling to police the privatized monopoly sufficiently, hence providing a problem of insufficient regulation (Virgin Atlantic 2007). This was but one of a litany of complaints by airlines that would ultimately push the regulators to move. In 2008, the UK Competition Commission announced that BAA would be compelled to sell several of its airports as a means of breaking its apparent territorial monopoly on London (owning and operating its three major airports—Heathrow, Gatwick, and Stansted).

What does this mean for conceptualizations of territoriality? First, work by mobilities theorists has highlighted the "seeming incommensurability between different versions of travel time" (Watts and Urry 2008, 860). Thus, while the apparently mundane issue of negotiating a prompt exit from a garage may not be high priority for public policy, the high price sensitivity of airlines makes this a key issue of profitability. This highlights the "disparity between travel-time use as valued and experienced by transport strategists and policy makers . . . and travel-time use as valued and experienced by passengers on-the-move" (860). Second, despite the association of connectivity pushed by politicians and globalization theorists alike, airports—with finite slot and passenger handling capacity—remain firmly within national regulatory frameworks. Third, it means that state governments (especially on a regional level) have to be more entrepreneurial in negotiating for their own connectivity, most noticeable in the growth of budget airline deals with provincial airports. There is now a powerful shadow geography of (apparently) budget airlines operating in (apparently) provincial regions. Closer examination may reveal, however, that these new airport–airline relationships reflect new geographies of mobility, based not on political pork barreling, but rather on new business models of competitively priced, no-frills short-haul flights.

Conclusions

> The airport seems to be pulling everything tighter together even as it seems to be pulling it all apart. It is a site where all kinds of transgressions across discrete spaces, times and systems occur—in which the very latest of air guidance control systems mingle with ancient wetlands and marine ecologies . . . Decentralized and deterritorialized, the city is rewritten according to a global logic of connection. (Fuller and Harley 2004, 106–7)

How does this discussion of the airport fit into broader debates about territory, flow, and urbanity? First, I have highlighted that airports operate at a scalar level that city—and even regional—planners find difficult to control. Major hubs are often seen as being of national importance, and central governments have rarely allowed environmental protest to hold up the expansion or construction of airports. As borderlands, airports are crucial sites in the regulation of mobility and are—usually—carefully sealed biometric regimes of surveillance (chapter 6). As with Schivelbusch's (1986) account of railway journeys in France, the airport gathers together landscapes of greater or lesser distance. To understand this is a methodological challenge, requiring a consideration of the multitude of travel modes that snake in and around airports, from the municipal service buses that connect the cargo depots, business parks, and terminals with worker housing, to the multispeed modes of connecting passengers to destinations (from maglevs through taxis and shuttle buses to the private cars that are left in satellite garages on the airport's outskirts), to the express exits offered by high-speed trains and connecting flights. Yet the operating requirements can be difficult to reconcile with human scale or fine-grained urban tissues, meaning that the "last mile" can be as crucial to an airport's success as its success in getting planes, passengers, and things in and out of its terminals. As a general rule, the smaller the airport the less time it takes to reach the desired final destination, given the traffic congestion and capacity issues facing the most popular airports such as Heathrow or Mumbai. The European Union's drive to complete a high-speed train network has highlighted the potential for intermodality (where short-haul flights from major hubs such as Paris, Frankfurt, or Amsterdam are replaced by high-speed trains that have a stop adjacent to airport terminals). So airports and air travel have to be understood within the social, material, and spatial practices of travel, where travel time per se is less important than travel time *use* (Watts and Urry 2008).

Second, airports are often presented as disembodied spaces, which construct "simplified and imaginary passengers with particular affective potentialities" (Adey 2008, 442). They tend to be seen as sites that allow for hypermobility, flicking a somewhat mythical "global elite" hither and thither across the world. The reality is a lot more complex, and it would be more appropriate to take the passenger arrival and destination statistics from individual airports to get a clearer idea of the intranational or transnational flows that predominate in each space. Only a relatively small number of airports could be genuinely labeled as "global," perhaps defined by an algorithm that measures airports with direct connections to the greatest number of nation-states, set against the citizenship origins of travelers, set against the sheer quantum of body flows that pass through their terminal buildings. It would also be appropriate to measure the financial attributes of these travelers, where the premium costs of first-class or, to a lesser extent, business-class tickets be measured as a proportion of the economy-class fees paid by travelers. A more differentiated account of the relationship between airports and airlines (which is one indicator of ticket prices) is required, as well as an internal accounting of the operating costs and balance sheets of airlines as firms. This, in turn, requires a consideration of the importance of comfort, now a core element within airline and airport marketing messages. These firms echo the marketing messages of railway companies that "enact a more complex and multifarious definition of comfort that emerges from a number of other sensibilities such as quietness, solitude, relaxation, slowness, and beauty" (Bissell 2008, 1701).

From here, there is a basis to consider how airport governance may in turn affect the embodied politics of everyday travel. For Bissell (2008), the growing scholarly interest in "sedentary affect" is evidence of a growing maturity in mobilities research, which now "is beginning to take seriously the relationality between stillness and movement and their mutual constitution" (1698). For the airport manager, this affords numerous commercial opportunities, including advertising and retail strategies aimed at temporarily immobile consumer-travelers (Adey 2008). These embodied cultural economies are calculable, and another reason why the specialisms of the globalized design and consultancy firms identified in the introduction are in such demand worldwide. Such firms have expertise in the logistics of processing, scanning, and comforting relatively slow movements of large numbers of people and their behaviors, reconciling this within an economic diagram of how the airport will work financially. With these caveats in mind, it is nonetheless fair to say that the airport can be seen as the machinic complex of globalization par excellence, a propeller that allows the fast projection of bodies and things over long distances at high speed, and at relatively low cost. As one of the earliest organizational forms that pursued international operational standardization, its operation has long enjoyed a high degree of predictability. This means that the airport is a format that is easily provisioned by globally organized consultancies, manufacturers, and systems operators, which would explain the tentative expansion into global airport management by companies such as Macquarie Airports and Schiphol Group. Unlike generic forms of property ownership, airports are an attractive investment for firms that operate strong sale and leaseback-style operations. While they do require substantial levels of capital investment to ensure their successful functioning, they are often embedded within predictable planning frameworks and may enjoy near monopolies of airport usage in their immediate territories.

The current landscape of airport development and management reveals much about the nature of urban governance within the contemporary nation-state. In part, this is due to the growing significance of the airport as a synecdoche of the "state of the nation," as an object of national pride, as the first impression of foreign visitors, as an important revenue generator and multiplier. Yet despite its obvious urbanity (its sense of being a totally manufactured and highly programmed environment, its high densities of users and labor, the raw expression of its dominance over surrounding countryside), the airport has an ambiguous relationship to the cities that it serves. The power of major airports in channeling and generating economic activity means that they actively rearticulate the spatial morphology of large metropolitan regions, strengthening the power of some economic corridors at the expense of other areas, generating important local labor markets, creating new markets for land development of a variety of uses, and—most controversially—introducing a third dimension of urban impact in the aerial noise corridors that massively affect local residential communities.

So airports can be seen as contradictory spaces within territorial development strategies. As time-space gateways, they are an important lubricant in the movement of people and things. But the intense and often competing interests of their stakeholders also make them difficult to manage and regulate. Airport management—which is increasingly privatized, with attendant "enclave" governance rights—is disruptive of broader planning regimes, unless the airport remains subordinate to broader state planning frameworks. And airport managers are also conscious that their primary stakeholders are not the "traveling public" but, rather, their globally distributed shareholders.

References

Adey, P. 2003. "Secured and Sorted Mobilities: Examples from the Airport." *Surveillance and Society* 1, no. 3: 500–19.

———. 2006. "Airports and Air-Mindedness: Spacing, Timing, and Using Liverpool Airport 1929–1939." *Social and Cultural Geography* 7, no. 3: 343–63.

———. 2008. "Airports, Mobility, and the Calculative Architecture of Affective Control." *Geoforum* 39:438–51.

Allen, J., and A. Cochrane. 2007. "Beyond the Territorial Fix: Regional Assemblages, Politics, and Power." *Regional Studies* 41, no. 9: 1161–75.

Amin, A., and N. Thrift. 2002. *Cities: Reimagining the Urban*. Cambridge: Polity.

Beauregard, R. A. 1993. *Voices of Decline: The Postwar Fate of U.S. Cities*. Cambridge, Mass.: Blackwell.

Beaverstock, J. V., R. G. Smith, and P. J. Taylor. 2000. "World-City Network: A New Metageography?" *Annals of the Association of American Geographers* 90, no. 1: 123–34.

Billig, M. 1995. *Banal Nationalism*. London: Sage.

Bishop, P. 2002. "Gathering the Land: The Alice Springs to Darwin Rail Corridor." *Environment and Planning D: Society and Space* 20:295–317.

Bissell, D. 2008. "Comfortable Bodies: Sedentary Affects." *Environment and Planning A* 40:1697–712.

Bouman, M. J. 1996. "Cities of the Plane: Airports in the Networked City." In *Building for Air Travel: Architecture and Design for Commercial Aviation*, edited by J. Zukowsky, 177–93. Munich: Prestel/Art Institute of Chicago.

Bruegmann, R. 1996. "Airport City." In *Building for Air Travel: Architecture and Design for Commercial Aviation*, edited by J. Zukowsky, 195–211. Munich: Prestel/Art Institute of Chicago.

Bunnell, T. 2004. *Malaysia, Modernity, and the Multimedia Super Corridor: A Critical Geography of Intelligent Landscapes*. London: RoutledgeCurzon.

Carroll, M., S. Vincent, J. Hassard, and F. L. Cooke. 2005. "The Strategic Management of Contracting in the Private Sector." In *Fragmenting Work: Blurring Organizational Boundaries and Disordering Hierarchies*, edited by M. Marchington, D. Grimshaw, J. Rubery, and H. Willmott, 89–109. Oxford: Oxford University Press.

Caruana, V., and C. Simmons. 2001. "The Development of Manchester Airport, 1938–78: Central Government Subsidy and Local Authority Management." *Journal of Transport Geography* 9:279–92.

Cresswell, T. 2006. *On the Move: Mobility in the Modern Western World*. London: Routledge.

Department for Transport. 2003. *The Future of Air Transport*. White paper. London: HMSO. http://www.dft.gov.uk/about/strategy/whitepapers/air/.

Doganis, R. 2006. *The Airline Business*. 2nd ed. London: Routledge.

Erie, S. P. 2004. *Globalizing LA: Trade, Infrastructure, and Regional Development*. Stanford, Calif.: Stanford University Press.

Fuller, G., and R. Harley. 2004. *Aviopolis: A Book about Airports*. London: Black Dog.

Goh, C. T. 2004. Keynote address. Presentation at the International Air Transport Association's 60th annual meeting and world air transport summit, June 7. http://www.iata.org/events/agm/2004/newsroom/keynote_address_pm.htm.

Graham, A. 2003. *Managing Airports: An International Perspective*. 2nd ed. Oxford: Butterworth-Heinemann/Elsevier.

Graham, S., and S. Marvin. 2001. *Splintering Urbanism: Networked Infrastructures, Technological Mobilities, and the Urban Condition*. London: Routledge.

Griggs, S., and D. Howarth. 2004. "A Transformative Political Campaign? The New Rhetoric of Protest against Airport Expansion in the UK." *Journal of Political Ideologies* 9, no. 2: 181–201.

Güller, M., and M. Güller. 2003. *From Airport to Airport City*. Barcelona: Gustavo Gili.

Hall, T., and P. Hall. 2006. "Heathrow: A Retirement Plan. Town and Country Planning Association, Tomorrow Series Paper 3." www.tcpa.org.uk/press_files/pressreleases_2006/Heathrow.pdf.

Humphries, I. 1999. "Privatisation and Commercialization: Changes in UK Airports." *Journal of Transport Geography* 7, no. 2: 121–44.

Jefferis, C., and F. Stilwell. 2007. "Private Finance for Public Infrastructure: The Case of Macquarie Bank." *Journal of Australian Political Economy* 58:44–61.

Kasarda, J. 2004. "Asia's Emerging Airport Cities." *Urban Land Asia*, December, 18–21.

———. 2006. "The Rise of the Aerotropolis. The Next American City." http://americancity.org/magazine/article/the-rise-of-the-aerotropolis-kasarda/.

Keeling, D. J. 1995. "Transport and the World City Paradigm." In *World Cities in a World-System*, edited by P. Knox and P. J. Taylor, 115–31. Cambridge: Cambridge University Press.

Marchington, M., D. Grimshaw, J. Rubery, and H. Willmott, eds., 2005. *Fragmenting Work: Blurring Organizational Boundaries and Disordering Hierarchies*. Oxford: Oxford University Press.

McLean, B. 2007. "Would You Buy a Bridge from This Man?" *Fortune*, October 1, 76–83.

O'Connor, K. 1995. "Airport Development in Southeast Asia." *Journal of Transport Geography* 3, no. 4: 269–79.

———. 2003. "Global Air Travel: Toward Concentration or Dispersal?" *Journal of Transport Geography* 11:83–92.

Olds, K. 2001. *Globalization and Urban Change: Capital, Culture, and Pacific Rim Mega-Projects.* Oxford: Oxford University Press.

Pile, S. 1999. "What Is a City?" In *City Worlds*, edited by D. Massey, J. Allen, and S. Pile, 3–52. London: Routledge.

Rimmer, P. J. 1991. "The Global Intelligence Corps and World Cities: Engineering Consultancies on the Move." In *Services and Metropolitan Development: International Perspectives*, edited by P. W. Daniels, 66–106. London: Routledge.

Salter, M., ed., 2008. *Politics at the Airport.* Minneapolis: University of Minnesota Press.

Schivelbusch, W. 1986. *The Railway Journey: The Industrialization of Time and Space in the 19th Century.* Berkeley: University of California Press.

Smith, D. A., and M. F. Timberlake. 2001. "World City Networks and Hierarchies, 1977–1997: An Empirical Analysis of Global Air Travel Links." *American Behavioral Scientist* 44:1656–78.

Timmons, H. 2006. "After Dubai Uproar, Emirates Air Holds No Grudges." *New York Times*, March 29. http://www.nytimes.com/2006/03/29/business/world business/29emirates.html.

Torrance, M. I. 2008. "Forging Local Governance? Urban Infrastructures as Networked Financial Products." *International Journal of Urban and Regional Research* 32, no. 1: 1–21.

Urry, J. 2000. *Sociology beyond Societies: Mobilities for the Twenty-First Century.* London: Routledge.

Virgin Atlantic. 2007. "Rewarding Failure: Airports Price Control Review—Initial Proposals for Heathrow, Gatwick and Stansted." Virgin Atlantic Response to CAA Consultation. https://www.caa.co.uk/docs/5/ergdocs/airportsfeb07/virgin.pdf.

Virilio, P. 1991. *The Lost Dimension.* New York: Semiotext(e).

Walters, W. 2002. "Mapping Schengenland: Denaturalizing the Border." *Environment and Planning D: Society and Space* 20:561–80.

Watts, L., and J. Urry. 2008. "Moving Methods, Travelling Times." *Environment and Planning D: Society and Space* 26:860–74.

Cities Assembled

Space, Neoliberalization,

(Re)territorialization, and Comparison

Kevin Ward and Eugene McCann

> The notion of urban assemblages in the plural form provides an
> adequate conceptual tool to grasp the city as a multiple object, to
> convey a sense of its multiple enactments.
>
> —Farias 2010, 14

> The complex of actors, powers, institutions, and bodies of
> knowledge that comprise expertise has come to play a crucial role
> in establishing the possibility and legitimacy of government.
>
> —Miller and Rose 2008, 69

This book has three goals. First, it endeavors to advance a theorization of urban policymaking and place making that understands both as
assemblages of "territorial" and "relational" geographies. It thus seeks
to demonstrate through theoretically informed empirical studies the folly
of polarized views in the current debate over theorizing space. Instead, it
argues for an approach that is sensitive to *both* the territorial and relational geographies that constitute the city. Second, it moves beyond the
limits of the traditional political science–dominated policy transfer literature, acknowledging its insights while also arguing for a broadening
of our understanding of agents of transference, a reconceiving of the
sociospatial elements of how policies are made mobile, and a departure
from the methodological nationalism and an acknowledgment of the

importance of interlocal policy mobility. Third, through the richness of its empirical examples, the book highlights the various methodological challenges that researchers face in attempting to operationalize an urban–global, territorial–relational conceptualization. Tracing the circuits, the networks, and the webs in and through which policies move from one site to another is not easy. It is also often not cheap, in financial terms, and in terms of the time it takes to do properly. That much is clear from the methodological insights generated from the likes of Burawoy (1998), Burawoy et al. (1991, 2000), and Marcus (1998), which demonstrate the complexities of linking "outcomes" and "processes" with various territorial units, such as the "local" and the "global."

Analyzing policies as they are territorialized, deterritorialized, and then reterritorialized, sometimes in a nonlinear, rhizomatic manner, demands careful exploration. Using qualitative techniques such as discourse analysis and semistructured interviews, many of the chapters in this collection seek to research geographically disparate but socially entangled and "proximate" sites. To discover the "where" of urban politics and policymaking, we must leave not just the confines of city hall but also the city itself, as Harvey (1989) puts it. Authors in this book take up Harvey's challenge to trace networks that encompass consultancy offices, firm headquarters, and think tank locations, as well as other sites such as conferences, seminars, and workshops.

Our view as editors is that these chapters constitute a serious intellectual attempt to uncover the various sites and techniques in and through which cities are *assembled*. "The concept of assemblage," for Wise (2005, 86), "shows us how institutions, organizations, bodies, practices and habits make and unmake each other, intersecting and transforming: creating territories and then unmaking them, deterritorializing, opening lines of flight as a possibility of any assemblage, but also shutting them down." By this we mean the ways in which cities, as "physical, biological, and social entities" are quite literally put together (DeLanda 2006, 16; see also Allen and Cochrane 2007; Prince 2010). Understanding cities in this way builds on a conceptualization of space as both territorial and relational or topological (Jones 2009) and to approach the sociospatial practices of urban policy actors in terms of both their fixities and their mobilities. For the two of us, this book marks not the ending, but the beginning of a sustained intellectual endeavor. And we are not alone. Over the past five years, a number of critical urban and regional scholars have been working away on some of the issues that have been discussed in the chapters of this book (Cook 2008; Guggenheim and Söderström 2010; McCann 2004, 2008, 2010; McCann and Ward 2009; Peck 2003;

Peck and Theodore 2001, 2009; Prince, 2010; Robinson 2008; Wacquant 1999; Ward 2006, 2007, 2010).

As with most edited collections, there is as much that divides as unites the individual contributions. From Massey's identification of a counter-hegemonic relationality of place (chapter 1) to Robinson's use of city strategies to document the ways in which spatial knowledge is circulated (chapter 2); from Peck's historical comparison of two "moments" of creativity in urban policymaking (chapter 3) to Ward's analysis of a policy in motion (chapter 4); from McCann's detailing of the various geographical reference points in the making of Vancouver's drug policy (chapter 5) to Keil and Ali's use of Severe Acute Respiratory Syndrome (SARS) to uncover the relational geographies of global cities (chapter 6); and finally to McNeill's study of airports as territorial outcomes of a nexus of sociospatial relationships (chapter 7), the chapters have used a range of methods and empirical cases to exemplify their individual arguments. As editors, we have not tried to overimpose a structure on the authors. Rather, we asked them to use their own theoretical reference points, deploying whatever research methods they thought most appropriate, to address the collection's three overarching goals. Some chapters are optimistic; their readings of the current urban condition are open to the possibility of progressive outcomes. Others are less so, emphasizing instead the limits to some aspects of current urban policymaking and remaining unconvinced about the gains being made by alternatives.

In concluding, and in synthesizing the collection's main findings, the rest of this chapter is organized into four sections. Each is a general theme that emerges from a number of the individual chapters. The first section turns to the current concern within human geography over how best to theorize space. This is a concern that troubles many of the contributors to this collection. It is also a debate in which some contributors have been centrally involved (Massey 1993, 1999, 2007; Robinson 2006). The second section turns to the relationship between the process of neoliberalization and the city. Again, the place of neoliberalization in the restructuring of cities, and the place of cities in the restructuring of neoliberalization is an issue that animates many of the individual contributions in this collection (see also Brenner et al. 2010a, 2010b; Leitner et al. 2007; Peck 2006). There is a degree of difference within the contributions. For some, the findings reinforce the centrality of neoliberalism in understanding the context within which much of politics and policymaking take place. For others, however, there is something else going on in the ways in which cities are positioned vis-à-vis each other. The combination of these perspectives in this volume suggests that references to neoliberalism cannot fully answer questions about contemporary urbanism and that, when these references are made, they must be undergirded by detailed empirical and theoretical analysis. The third section examines the territorializing and reterritorializing tendencies underpinning contemporary policymaking and place-making activities. A number of chapters carefully trace the ways in which policies are uprooted, mobilized, and circulated across space, transformed in some cases along the way. They also highlight how such policies are reterritorialized or embedded in particular contexts (see also McCann 2011; McCann and Ward 2010; Ward 2006, 2007, 2010). The fourth section turns to the comparative logic that underscores much of contemporary urban policymaking. Emphasizing the role comparison plays in governance at a distance, various chapters argue that an increasingly benchmarked society renders the previously unknowable knowable, the previously incomparable comparable (Larner and Le Heron 2002, 2004; Robinson 2006; Ward 2008).

In conclusion, and in drawing together the empirical and theoretical insights of the individual chapters, we make two points. First, that thinking through the territorial and relational geographies at work in the constitution of the contemporary urban condition remains a fruitful area for future theoretical work. Cities are implicated in each other's futures. As Hart (2002a, 91; original emphasis) argues most forcefully, there is a need to come "to grips with persistently diverse but increasingly *interconnected* trajectories of sociospatial change in different parts of the world." Second, in searching for the whereabouts of "urban" politics and policy, scholars need to be attuned to the variety of potential sites—from hotel lobbies to conference centers, city halls to corporate headquarters, public parks to school playgrounds—in which knowledge is mobilized and struggled over. This picks up on earlier insights from Harvey (1989) and stretches the point in light of some more recent insights from poststructural human geographers who have argued for a more-than-territorial conceptualization of the city (see Jones 2009). Our contention is that this volume has, in its own way, made a modest contribution to this still fledgling multidisciplinary intellectual field.

Spaces as Both Territorial and Relational

In recent years, it has become something of a truism to argue for an understanding of cities that is both territorial and relational. Or, put another way, for a theorization that, topologically speaking, is alive to "the multiple spatial networks that any city is embroiled in, and to . . . the full force of these networks and their juxtaposition in a given city upon local dynamics" (Amin 2007, 112). A series of influential contributions have slowly but surely sought to place *thinking relationally* at the core of contemporary human geographic scholarship. What in the early 1990s started slowly has however more recently generated a degree of pace. Over the past ten years or so, in particular, this approach to theorizing space has become almost de rigueur, especially among many urban economic and political geographers (Allen et al. 1998; Allen and Cochrane 2007, 2011; Hart 2002a, 2002b; Jones 2009; Jones and MacLeod 2004; MacLeod and Jones 2007; McCann and Ward 2010; Massey 1993, 1999, 2007; Morgan 2007). It is captured by Hart (2002a, 297) who argues that "[cities should be understood as] dense bundles of social relations and power-infused interactions that are always formed out of entanglements and connections with dynamics at work in other places, and in wider regional, national and transnational arenas." And Jones (2009, 2) goes so far as to state that "thinking space relationally . . . is becoming the mantra of early twenty first century human geography." What has emerged from this creative dialogue is an intellectual arena in which those working on a whole series of issues have coalesced around the need to theorize the city in a very particular way. That is, to understand the city as *both* open, internally heterogeneous, constituted through its place in myriad of connections and networks *and* as a territorially institutionalized object, the outcome of various "political" contestations and struggles. This view is neither about junking entirely territorial notions of the city nor about embracing uncritically understandings of them solely as "places of nodal connectivity, inflected by the overlaps of historical legacy and spatial contiguity" (Amin 2007, 112). Rather, it is one that understands circuits, networks, and webs as profoundly spatial, *and* scales and territories as dynamic and constantly in the process of emerging, disappearing and reemerging (Jones and MacLeod 2004; Jones 2009). It is this view of the city as "both a place (a site or territory) and as a series of unbounded, relatively disconnected and dispersed, perhaps sprawling and differentiated activities, made in and through many different kinds of networks stretching far beyond the physical extent of the city" (Robinson 2006, 121) that underscores the contributions in this collection.

Building on her earlier work, Massey (chapter 1) is clear that she understands space as relational. Yet she rejects the extreme versions of the relational position that might argue that, in the current global era, there is no fixity or territoriality; only motion, flow, and connections. Her view holds that cities exist within geographies of economic, cultural, political relations and that places, as territorial expressions, are constituted through these. This is a view held by others in this volume. Robinson (chapter 2), McCann (chapter 5), Keil and Ali (chapter 6), and McNeill (chapter 7) all outline their understandings of space, both through their respective theoretical frameworks and through the ways in which they use their empirical work. Keil and Ali appreciate the continuing importance of bounded territories but argue for an understanding of the city that reflects how the global SARS crisis transcended continents, city-states, and communities, each with their own variegated political systems and social and cultural ecologies. The chapters by Peck and Ward also seek to highlight how cities are topological. However, both also emphasize, perhaps more than the other chapters, the continuing importance of cities as territorial configurations.

Neoliberal Urbanization

The collection's second theme involves neoliberal urbanization. Over the past decade a large amount of work has been produced from across the social sciences on neoliberalism and the city (Brenner and Theodore 2002; Brenner et al. 2010a, 2010b; Hackworth 2007; Peck 2006; Ward 2007; Wilson 2004). This is, of course, just one element in a much larger intellectual field. Human geographers of many different theoretical persuasions have set about analyzing "neoliberalism," along the way playing a role in constructing their very object of investigation. Whether working on substantive areas such as crime, education, or the environment, neoliberalism has been "understood variously as a bundle of (favored) policies, as a tendential process of institutional transformation, as an emergent form of subjectivity, as a reflection of realigned hegemonic interests, or as some combination of the latter" (Brenner et al. 2010b, 183).

There is now a significant body of human geographical scholarship on what we might think of as neoliberal urbanization, and the notion of neoliberalization is central to several of the chapters in this volume. Their analyses often mirror that of Brenner et al. (2010b, 215), who highlight "the path-dependent interactions between neoliberal projects of restructuring and inherited institutional and spatial landscapes . . . [and] . . . emphasize the geographically variable, yet multiscalar and translocally interconnected, nature of neoliberal urbanism." In terms of the wider urban system, the many different variants of neoliberalism have shaped how different cities within the global urban system have experienced its restructuring tendencies and how individual cities have experienced these tendencies in various ways:

> Cities have served as command centers, relay stations, and experimental sites in the roll-out of neoliberal modes of governance: they have

been battlegrounds in strategically significant moments in the process of neoliberalization; and they have been epicenters of contestation and transformative struggle. (Leitner et al. 2007, 315)

Massey's chapter situates contemporary London as an outcome of neoliberal tendencies, particularly around the global restructuring of the finance industry. The growth of the city of London has put it at the center of all kinds of relations—of investment, trade, security, insurance, speculation—many "regressive" but some "progressive." What is clear is that that there are many different Londons, all situated differently in relation to the production of the current form of neoliberal globalization. Through this account, Massey sees some hope. She highlights the progressive possibilities of using the connections that bind cities together in a positive and progressive manner. Robinson picks up on the point about neoliberalization's production. City Development Strategies have been seen by some as disempowering to "third world" cities, forcing them down the broadly neoliberal development agenda so liked by many international agencies. Robinson holds out more hope, however. Acknowledging the unequal power relations within and among cities, she argues that the potential remains for city strategies to act as a means through which urban futures other than neoliberal ones might be imagined. Similarly, in McCann's discussion, the sorts of knowledge-sharing practices (conferencing, site visits, Internet connections, etc.) that are so commonly discussed in terms of the neoliberal "scanning" of the global policy landscape are shown to have shaped an innovative drug policy in Vancouver, which is, according to its scientific evaluators, saving lives among the city's poorest, most marginal population.

The two chapters by Peck and by Ward both seek to uncover the means through which neoliberal urbanization comes about. For the former, the focus is on the production and the circulation of notions of "creative cities," while for the latter, the analysis of the international emergence of Business Improvement Districts also speaks to current fads in economic development. Use of terms such as *business climate* and *quality of life* punctuate both sets of circulations, and in many cases, the language of "creativity" is used by Business Improvement Districts as they seek to remake urban downtowns from around the world. In both these chapters, resistance to neoliberalism is absent from the accounts. This is not because there is none. There is but not much. Rather, creative city policies developed and delivered through growth coalitions involving Business Improvement Districts have become increasingly normalized for those working on urban political issues in many cities. McNeill picks up the relationship between neoliberalization and norms around economic development strategies. The language of competitiveness and global connectivity has been used regularly to explain and to justify the increased attention paid to airports—the subject of his chapter—as drivers of urban globalness.

These different perspectives on the current neoliberalized urban context suggest that neoliberal circuits of knowledge and norms of practice might be used for a range of political and policymaking purposes, rather than simply in aid of the sorts of neoliberalization we are used to thinking about (Larner 2003). In other words, we suggest that the chapters in this volume show that the contemporary territorial–relational complex that produces cities is about neoliberalism but is also about *much more* than neoliberalism in its narrowly defined sense.

Deterritorializing and Reterritorializing Policies

The third theme that runs through many of this collection's chapters is that of mobile urban policies. Whether in the area of creativity (Peck, chapter 3), drugs (McCann, chapter 5), or the downtown (Ward, chapter 4), these chapters provide empirical evidence of deterritorializing and reterritorializing tendencies at play in the world of contemporary policy and place making. Whether formal guidelines (policies), statements of ideal policies (policy models), or expertise in the implementation and evaluation of policy (policy knowledge), there is a politics to their mobilization (and immobilization). Various agents are involved at different stages, some with clearer territorial attachments than others, some with longer geographical reaches than others, and often spurred by previous failures:

> Policy failure is central to the exploratory and experimental modus operandi of neoliberalization processes—it is an important impetus for their continual reinvention and ever-widening interspatial circulation. Indeed, rather than causing market-oriented regulatory projects to be abandoned, endemic policy failure has tended to spur further rounds of reform within broadly neoliberalized political and institutional parameters. (Brenner et al. 2010b, 209)

Critically engaging with work in political science on policy transfer (Wolman 1992; Dolowitz and Marsh 2000; Wolman and Page 2000, 2002; Stone 1999), this volume builds on a series of recent studies into the infrastructures that are both the consequence of, and facilitators for, the city-to-city movement of policies. This argues that a range of transfer agents, including those within the formal structures of the state, private policy consultants, think tank officers, academics, and activist groups, are involved in an ongoing process of identification, learning, and, in some cases, adopting policies

(Theodore and Peck 2000; Peck and Theodore 2001, 2009; Peck 2003, 2006; McCann 2011; Hoyt 2006; Ward 2006; Cook 2008).

Robinson (chapter 2) argues that city strategies are most usefully assessed through a properly global and transnational analytical focus, assessing them as a global urban policy technique, whose circulations and meanings extend beyond one region or category of cities. Using the term *spaces of circulation*, she states that understanding the mobility of urban policy necessitates attention to the sites and the tracks that compose a globalizing field of urban development knowledge. Her conclusion is upbeat. Robinson suggests that thinking through the uneven power-geometries of global urban governmentality reveals that city strategies can act as sites of opportunity for envisioning alternative urban futures. Not always quite so upbeat, perhaps, but working off the same theoretical page are Ward (chapter 4) and McCann (chapter 5). The latter uses the example of the making of drug policy in Vancouver to illustrate how points of reference, beyond the immediate territorial context, can become embedded in local political debate through travel, representation, repetition, and contest. His account draws attention to the genuine improvements in many drug users' lives wrought by importing harm reduction policies into Vancouver. Ward, however, uses the example of the Business Improvement District model of downtown governance to reveal the ways in which "examples" are made into "models" that can then be rendered mobile. Both chapters seek to draw together a range of theoretical literatures as a means of moving toward an approach that understands the politics of policy transfer as both relational and territorial. Together with Peck's chapter 3, which compares in a heterodox manner two periods in which "creativity" was used as a means of instigating economic development, McCann's and Ward's chapters speak to the emerging literature on the motion, movement, and transfer of policy. This is a point picked up by McNeill (chapter 7). The global growth in the number of airports is in no small part the result of the emergence of neoliberal models of economic development. Facilitated by international consultants, the expansion of airports reflects how investing in infrastructure, particularly around transport, has become bound up with city marketing strategies.

Governance at a Distance: Benchmarking, Comparison, and the Construction of a Global Knowledge Infrastructure

The fourth theme that runs through many of the chapters in this book is the role of comparison in shaping the contemporary city. "Comparison" has evolved into a central feature of economic development orthodoxy

(Peck 2003; McCann 2004; Ward 2008). Cities are compared against one another in all manner of ways, according to a raft of different indicators. For this to be possible, a series of technologies have been put in place. Or put another way, what Robinson (chapter 2), Ward (chapter 4), and Peck (chapter 3) highlight are methods through which "rules" of interlocal competition are constituted. The empirical cases discussed in these chapters, together with other studies, reveal that, while the techniques may have had their origins in the firm, international comparative techniques, such as benchmarking, are now just as likely to be practiced in the public sector and associated arm's-length agencies. State-sponsored transnational or transurban networks of policymakers and practitioners orchestrated around the disciplining consequences of comparison are increasingly common occurrences. One of the tendencies in the restructuring of many Western nation-states identified by Jessop (2002, 201), alongside the *destatization* and *denationalization of the state,* is the *internationalization of policy regimes,* by which he means the "development of the interregional and cross-border linkages connecting local and regional authorities and governance regimes in different national formations."

The chapters by Robinson, Ward, Peck and McCann would appear to reaffirm Larner and Le Heron's (2004, 213) argument that "benchmarking, along with audit and contractualism is a key technique of neo-liberalism," benchmarking being about comparison, in contrast to auditing (about checking) and contractualism (about performing). Indeed, Larner and Le Heron (2004, 212; our emphasis) argue that "comparative quantitative techniques such as indicators, standards and benchmarks *now play a central role in constituting globalizing economic spaces.*" The ways in which these four chapters speak to the disciplining role of these seemingly "mundane" practices reveal how quite ordinary and technocratic techniques require careful analysis. Under certain conditions, they can obfuscate the uneven geographical relations that bind some cities together and push others apart. This theme is also present in Keil and Ali (chapter 6) and McNeill (chapter 7). Cities are often ranked according to criteria that include quality of transport, or connectivity. Having a major hub airport near a city is seen as a plus and will move the city higher up the league table. Of course, there is a delicate balance to be maintained: the airport should not be too close, lest it affects air quality, and it should not be too far away lest it makes travel times unattractive. Furthermore, as Keil and Ali show, having a globally connected airport exposes cities to emerging infectious disease. From the headline-grabbing panics around the spread of infectious disease to the mundane practices of managing downtown business districts, these case studies and the others in this volume suggest a need for continued and heightened attention to the ways in which cities across the globe are positioned in relation to various "elsewheres" while also being assembled in particular territorial contexts.

Conclusion

This sort of positioning is evident in *Bombay First*, an initiative to make this particular Indian city "a better place to live, work and invest in" (http://www.bombayfirst.org). It was established "to act largely as a think tank of the city, and to assume a more specific role of forming partnerships between the Government, the Private Sector and the Civil Society" according to its Chair, N. K. Nayar (http://www.bombayfirst.org/bombay_first_original.html). It aims to serve the city with the best that private business can offer. It will achieve its goal "by addressing the problems of today and the opportunities of tomorrow, through partnerships with government, business and civil society." *Bombay First* commissioned the global consultancy firm McKinsey to produce a study entitled *Vision Mumbai: Transforming Mumbai into a World Class City*, published in 2003. It was well received among the city's growth coalition but received mixed reviews in the city more widely and in the surrounding region. Nevertheless, the initiative has played an important role in influencing policymakers both at the central and state levels to recognize and address the immediate need for urban renewal in the city. *Bombay First* intended to transform "Mumbai into a world class city with a vibrant economy and globally comparable quality of life for its citizens" (Vision Mumbai 2003, n.p.).

This policy document exemplifies the themes highlighted in this volume. Great stock is placed on restructuring the state and on making it even more involved in the activities of the private sector. Enterprise is to be liberated, but only under certain conditions and only if channeled to the right ends. The conditions of the poorest in society are best attended to through the public and private sectors working in unison, so it is argued. So there is some evidence of neoliberalism, perhaps? Of course, there is much that goes on in the neoliberal urban condition that is not neoliberal, as we have suggested previously. The document consists of progressive as well as regressive possibilities. Civil society groups, as well as a number of elite actors, have participated in its production. Indeed, its production created a political space in which various stakeholders were able to articulate a series of visions for the city. Some were taken more seriously than others. The ownership of the document is thus not straightforward. It promises to improve the lot of all of Bombay's residents, and thus it should not be dismissed as *just* another example of neoliberalism "as usual."

Of course, Mumbai is not the first city to produce this sort of "city strategy," as Robinson's chapter suggests. Nor will it be the last. Overlapping with the channels in and through which "creative" urban policies have diffused around the world, visioning documents of a self-consciously reflexive strategic nature have become standard fare for many cities since at least the 1990s (see chapter 3). Comparisons of various sorts are also present in *Vision Mumbai*, wherein Mumbai is situated next to and compared with Shanghai and Singapore most noticeably. These are places—success stories—to which Mumbai and those who govern it should aspire. So the initiative was shaped by comparisons among Mumbai and other cities *in* the global South both in terms of the cities' internal, "territorial" characteristics and in terms of how they are positioned (and aspire to be positioned) among wider interlocal relations. It was also shaped by attempts to "benchmark" Mumbai against a range of criteria, some of which speak to the understanding of cities as nodes, as places to move through as well as to be resident in (see chapter 7). Underpinning the diagnostic built into *Vision Mumbai* is a particular view on what it means to be a global city.

We think this vignette is revealing. Many of the issues that are the focus of this volume are present in it. While *Bombay First* spent a number of months after the publication of the document in negotiations with other local stakeholders about the future of the city, McKinsey, the consultancy firm, moved on to its next urban government client. No doubt the consultants drew on their experiences in producing *Vision Mumbai*. In the current economic climate, there is always another Mumbai, a city wanting to commission one or more agents of transference to produce some sort of visioning document. Peck's chapter 3 reveals how, in the current era, policies are quicker to circulate through practitioner and policymaking channels. The world of policymaking does appear to be one that never sleeps: doze off and someone else, or somewhere else, might overtake you. At least that is often how it appears. As Larner and Le Heron (2004, 215) argue about the technique of benchmarking, it "encourages places and people to constantly reinvent themselves and remobilize their efforts." Robinson (chapter 2), Peck (chapter 3), Ward (chapter 4), and McCann (chapter 5) provide evidence of the kinds of systems needed to facilitate this high-speed policy mobility. New policies from faraway places are introduced in the local political arena. Evidence is marshaled to legitimize the process. Contemporary urban policymaking, at and across all geographical scales, therefore involves monitoring the policy landscape, via trade publications and reports, the media, Web sites, blogs, professional contacts, and word of mouth. The search is for ready-made, off-the-shelf policies and best practices that can be quickly applied locally.

In conclusion, we wish to make three points. First, in different ways and to different degrees all the authors in this volume have sought to develop an approach to their own area of study that is sensitized to both territorial and relational conceptions of space. The unique experiences of the city, whether London (Massey), Vancouver (McCann), Detroit (Peck), or Singapore (Keil and Ali), have been analyzed through examining the wider relationships in which each is embedded. Whether the trade flows that bound together Caracas and London (Massey), or the viral flows of SARS that impacted Hong Kong, Singapore, and Toronto and precipitated further flows of knowledge between those cities and the World Health Organization's headquarters in Geneva (Keil and Ali), the contributors have also detailed empirically the ways in which the city is assembled through connections and relationships that have both short and long geographical reach. The authors have demonstrated an "expanded geographical imagination" of the sort urged by Amin (2007, 112). They have also continued to theorize cities as social formations, in and through which "territory" and "territorialization" as ontological, political, and policy-relevant categories remain important. So whether it is the construction and regulation of a new airport (McNeill, chapter 7) or the redesigning of a downtown (Ward, chapter 4), policy and politics happen *somewhere*. Territory and the various claims over it continue to matter, as the movement and transfer of policy necessarily involve encountering spatial forms that are preexisting and works in progress.

Second, the authors in this collection have demonstrated the range of spaces involved in the assembly of urban policies and urban place-making activities. In addition to those that traditionally have figured in accounts, particularly the city hall, a range of other sites have appeared as important. Particularly events, seminars, and workshops, what we might think of as spaces of learning, have featured prominently in the accounts by Robinson (chapter 2), Ward (chapter 4), Peck (chapter 3), and McCann (chapter 5). From ways of dealing with SARS (Keil and Ali, chapter 7) to how to make cities more creative (Peck, chapter 3), from how to construct Key Performance Indicators (Ward, chapter 4) to how to organize the interior of an airport (McNeill, chapter 7), a number of chapters have drawn attention to the importance of what at first glance are mundane activities and practices.

Third, this collection has revealed quite fruitfully some of the ways in which state rescaling has had important consequences for how cities of different types are assembled and disassembled, for how different cities are governed, and for how people live differently in different cities. It is clear that restructuring the state at all levels and reordering linkages among state agencies, private business (including private policy experts), and communities of various forms have affected both on the "internal" character of cities and on the "external" linkages among cities. This perhaps will come as no surprise to those readers au fait with the voluminous literature on the state produced by social scientists in the last two decades (Brenner 2004; Brenner et al. 2003; MacLeod and Goodwin 1999; Jessop 2002; Jones 1998; Jones and Ward 2002; Peck 2001). Swyngedouw (2005, 1992) writes about governing-beyond-the-state, by which he means a "much greater role in policy-making, administration and implementation to private economic actors on the one hand, and to parts of civil society on the other in self-managing what until recently was provided or organized by the national or local state." The new state forms have been neatly captured by the term *neoliberalization*, a concept with significant analytical purchase. Accepting the variegated nature of neoliberalization (Brenner et al. 2010a, 2010b), allows for its understanding as simultaneously patterned and regularized *and* contextual and hybridized. Doing so also holds it open, as a cultural, economic, and political system in which there is a room for agency, for maneuver, and for imagining a different urban future. For the technologies that allow policies to be moved from one city to another also work in the favor of activist and civil society movements. Groups seeking to challenge the neoliberal urban condition go on study tours, exchange "best practices" over the Internet, and seek to challenge "local" decisions through reference to elsewhere (Massey, chapter 1; Robinson, chapter 2; and McCann, chapter 5, most explicitly acknowledge how this might work). Likewise, comparisons may also hold emancipatory potential, as a means of naming and shaming unethical or unjust city authorities, for example. That examples of these "alternatives" remain still rare is no reason—theoretically, empirically, methodologically, or politically—to discount them. Rather the intellectual project that this volume is part of has as its objective to provide a fuller account of the processes and potentialities of interurban connection.

References

Allen, J., and A. Cochrane. 2007. "Beyond the Territorial Fix: Regional Assemblages, Politics, and Power." *Regional Studies* 41:1161–75.

Allen, J., and A. Cochrane. 2011. "Assemblages of State Power: Topological Shifts in the Organization of Government and Authority." *Antipode* 43.

Allen, J., D. Massey, and A. Cochrane. 1998. *Re-Thinking the Region*. London: Routledge.

Amin, A. 2007. "Re-Thinking the Urban Social." *City* 11:100–114.

Brenner, N. 2004. *New State Spaces: Urban Governance and the Rescaling of Statehood*. Oxford: Oxford University Press.

Brenner, N., B. Jessop, M. Jones, and G. MacLeod. 2003. *State/Space: A Reader*. Oxford: Blackwell.

Brenner, N., J. Peck, and N. Theodore. 2010a. "After Neoliberalization." *Globalizations* 7: 327–435.

Brenner, N., J. Peck, and N. Theodore. 2010b. "Variegated Neoliberalization: Geographies, Modalities, Pathways." *Global Networks* 10:182–222.

Brenner, N., and N. Theodore, eds. 2002. *Spaces of Neoliberalism: Urban Restructuring in North America and Western Europe*. London: Blackwell.

Burawoy, M. 1998. "The Extended Case Method." *Sociological Theory* 16:4–33.

Burawoy, M. 2000. "Introduction: Reaching for the Global." In *Ethnography Unbound: Power and Resistance in the Modern Metropolis*, edited by M. Burawoy, J. A. Blum, S. George, Z. Gille, T. Gowan, L. Haney, M. Klawiter, S. H. Lopez, S. Ó Riain, and M. Thayer, 1–40. Berkeley: University of California Press.

Burawoy, M., A. Burton, A. Arnett Ferguson, K. J. Fox, J. Gamson, N. Gartrell, L. Hurst et al. 1991. *Ethnography Unbound: Power and Resistance in the Modern Metropolis*. Berkeley: University of California Press.

Cook, I. R. 2008. "Mobilising Urban Policies: The Policy Transfer of U.S. Business Improvement Districts to England and Wales." *Urban Studies* 444:773–95.

De Landa, M. 2006. *A New Philosophy of Society: Assemblage Theory and Social Complexity*. New York: Continuum.

Dolowitz, D., and D. Marsh. 2000. "Learning from Abroad: The Role of Policy Transfer in Contemporary Policy Making." *Governance* 13:5–24.

Farias, I. 2010. "Introduction: Decentering the Object of Urban Studies." In *Urban Assemblage: How Actor Network Theory Changes Urban Studies*, edited by I. Farias and T. Bender, 1–24. New York: Routledge.

Guggenheim, M., and O. Söderström. 2010. *Re-Shaping Cities: How Global Mobility Transforms Architecture and Urban Form*. New York: Routledge.

Hackworth, J. 2007. *The Neoliberal City*. Ithaca, N.Y.: Cornell University Press.

Hart, G. 2002a. *Disabling Globalization: Places of Power in Postapartheid South Africa*. Berkeley: University of California Press.

Hart, G. 2002b. "Geography and Development. Development/s beyond Neoliberalism: Power, Culture, Political Economy." *Progress in Human Geography* 26:812–22.

Harvey, D. 1989. "From Managerialism to Entrepreneurialism—The Transformation in Urban Governance in Late Capitalism." *Geografiska Annaler Series B* 71:3–17.

Hoyt, L. 2006. "Importing Ideas: The Transnational Transfer of Urban Revitalization Policy." *International Journal of Public Administration* 29:221–43.

Jessop, B. 2002. *The Future of the Capitalist State.* Cambridge: Polity Press.

Jones, M. 1998. "Restructuring the Local State: Economic Governance or Social Regulation?" *Political Geography* 17:959–88.

———. 2009. "Phase Space: Geography, Relational Thinking, and Beyond." *Progress in Human Geography* 33, no. 4: 487–506.

Jones, M., and G. MacLeod. 2004. "Regional Spaces, Spaces of Regionalism: Territory, Insurgent Politics, and the English Question." *Transactions of the Institute of British Geographers* 29:433–52.

Jones, M., and K. Ward. 2002. "Excavating the Logic of British Urban Policy: Neoliberalism as the Crisis of Crisis-Management." *Antipode* 34:473–94.

Larner, W. 2003. "Guest Editorial: Neoliberalism?" *Environment and Planning D: Society and Space* 21:508–12.

Larner, W., and R. Le Heron. 2002. "The Spaces and Subjects of a Globalizing Economy: A Situated Exploration of Method." *Environment and Planning D: Society and Space* 20:753–74.

———. 2004. "Global Benchmarking: Participating 'at a Distance' in the Globalizing Economy." In *Global Governmentality: Governing International Spaces,* edited by W. Larner and W. Walters, 212–32. New York: Routledge.

Leitner, H., E. Sheppard, and J. Peck. "Squaring Up to Neoliberalism." In *Contesting Neoliberalism: Urban Frontiers,* edited by H. Leitner, E. Sheppard, and J. Peck, 311–27. New York: Guilford Press.

MacLeod, G., and M. Goodwin. 1999. "Space, Scale, and State Strategy: Rethinking Urban and Regional Governance." *Progress in Human Geography* 23:503–27.

MacLeod, G., and M. Jones. 2007. "Territorial, Scalar, Networked, Connected: In What Sense a 'Regional World'?" *Regional Studies* 41:1177–91.

Marcus, G. 1998. *Ethnography through Thick and Thin.* Princeton, N.J.: Princeton University.

Massey, D. 1993. "Power-Geometry and a Progressive Sense of Place." In *Mapping the Futures: Local Cultures, Global Change,* edited by J. Bird, B. Curtis, T. Putman, G. Robertson, and L. Tickner, 59–69. New York: Routledge.

———. 1999. "Imagining Globalisation: Power-Geometries of Space-Time." In *Global Futures: Migration, Environment, and Globalisation,* edited by A. Brah, M. Hickman, and M. MacanGhaill, 27–44. Basingstoke, UK: St. Martin's Press.

———. 2007. *World City.* Cambridge: Polity Press.

McCann, E. J. 2004. "'Best Places': Interurban Competition, Quality of Life, and Popular Media Discourse." *Urban Studies* 41:1909–29.

———. 2008. "Expertise, Truth, and Urban Policy Mobilities: Global Circuits of Knowledge in the Development of Vancouver, Canada's 'Four Pillar' Drug Strategy." *Environment and Planning A* 40:885–904.

———. 2011. "Urban Policy Mobilities and Global Circuits of Knowledge: Toward a Research Agenda." *Annals of the Association of American Geographers* 101.

McCann, E. J., and K. Ward. 2010. "Relationality/Territoriality: Toward a Conceptualization of Cities in the World." *Geoforum* 41:175–84.

Miller, P., and N. Rose. 2008. *Governing the Present.* London: Polity Press.

Morgan, K. 2007. "The Polycentric State: New Spaces of Empowerment and Engagement?" *Regional Studies* 41:1237–51.

Peck, J. 2001. *Workfare States.* New York: Guildford Press.

———. 2003. "Geography and Public Policy: Mapping the Penal State." *Progress in Human Geography* 27:222–32.

———. 2006. "Liberating the City: Between New York and New Orleans." *Urban Geography* 27:681–723.

Peck, J., and N. Theodore. 2001. "Exporting Workfare/Importing Welfare-to-Work: Exploring the Politics of Third Way Policy Transfer." *Political Geography* 20:427–60.

———. 2009. "Embedding Policy Mobilities." Working paper, Department of Geography, University of British Columbia.

Prince, R. 2010. "Policy Transfer as Policy Assemblage: Making Policy for the Creative Industries in New Zealand." *Environment and Planning A* 42:169–86.

Robinson, J. 2006. *Ordinary Cities: Between Modernity and Development.* London: Routledge.

———. 2008. "Developing Ordinary Cities: City Visioning Processes in Durban and Johannesburg." *Environment and Planning A* 40:74–87.

Stone, D. 1999. "Learning Lessons and Transferring Policy across Time, Space, and Disciplines." *Politics* 19:51–59.

Swyngedouw, E. 2005. "Governance Innovation and the Citizen: The Janus Face of Governance-beyond-the-State." *Urban Studies* 42:1991–2006.

Theodore, N., and J. Peck. 2000. "Searching for Best Practice in Welfare-to-Work: The Means, the Method, and the Message." *Policy and Politics* 29, no. 1: 81–98.

Wacquant, L. 1999. "How Penal Common Sense Comes to Europeans: Notes on the Transatlantic Diffusion of the Neoliberal Doxa." *European Societies* 1:319–52.

Ward, K. 2006. "'Policies in Motion,' Urban Management, and State Restructuring: The Trans-Local Expansion of Business Improvement Districts." *International Journal of Urban and Regional Research* 30:54–75.

———. 2007. "Business Improvement Districts. Policy Origins: Mobile Policies and Urban Liveability." *Geography Compass* 2:657–72.

———. 2008. "Editorial—Towards a Comparative (Re)Turn in Urban Studies? Some Reflections." *Urban Geography* 28:405–10.

———. 2010. "Entrepreneurial Urbanism and Business Improvement Districts in the State of Wisconsin: A Cosmopolitan Critique." *Annals of the Association of American Geographers* 100.

Wilson, D. 2004. "Towards a Contingent Neoliberalism." *Urban Geography* 25:771–83.

Wise, J. M. 2005. "Assemblage." In *Gilles Deleuze: Key Concepts*, edited by C. J. Stivale, 77–87. Durham, UK: Acumen Publishing.

Wolman, H. 1992. "Understanding Cross-National Policy Transfers: The Case of Britain and the U.S." *Governance: An International Journal of Policy and Administration* 5:27–45.

Wolman, H., and E. Page. 2000. *Learning from the Experience of Others: Policy Transfer among Local Regeneration Partnerships*. York: Joseph Rowntree Foundation.

———. 2002. "Policy Transfer among Local Government: An Information-Theory Approach." *Governance* 15, no. 4: 477–501.

ACKNOWLEDGMENTS

The thinking behind this volume took place toward the end of 2006. At that time, our work was leading us to ask similar sorts of questions about contemporary cities and their relationships with one another. A number of meetings at Association of American Geographers and Royal Geographical Society-Institute of British Geographers annual conferences afforded us opportunities to talk issues through. One of us thought that editing a book would be relatively painless. The other one believed him, which makes both of us naïve! Nevertheless, editing this book has cemented our friendship. Like all good working relationships, ours has involved a healthy combination of scholarship and socializing, of alcohol and food. And we're still talking to one another at the end of it.

In the production of this volume, we have accrued some intellectual debts, and it is time to acknowledge these. Thank you to the referees who passed comment on the proposal. Together with the contributors, we have done our best to attend to your concerns. At the University of Minnesota Press, Pieter Martin has been very supportive, overseeing the smooth production of the book with a minimum of fuss. Rini Sumartojo was also a great help in finalizing the manuscript. Thanks also to Danny Constantino for copyediting, Sallie Steele for indexing, and to the Simon Fraser University Publications Committee for funding the indexing. Allan Cochrane was good enough to write the foreword. His work has been hugely influential to our thinking, and we were delighted that he agreed to be involved. We'll always be grateful to the two academic editors, Susan Clarke and the late Gary Gaile, for commissioning this volume. We've dedicated the book to Gary's memory. His humor was sorely missed as we brought the volume together.

Kevin also acknowledges a 2005 Philip Leverhulme Prize. Closer to our homes, our debts largely lie with our respective family members. For Eugene, thanks go to Chris and to their daughter, Sophie, as well as to his mother, Maureen, and late father, James, to whose memory the book is also dedicated. For Kevin, thanks go to Colette and to their son, Jack.

Eugene McCann, Burnaby, Canada
Kevin Ward, Manchester, England

Allan Cochrane is professor of urban studies at the Open University. His research interests lie at the junction of geography and public policy, and he has researched and published on a wide range of topics relating to urban and regional policy. He is particularly interested in exploring the connections between policy change, institutional restructuring, and wider processes of political and social change. Allan has recently been undertaking research on reshaping and reimagining Berlin, on which he has published extensively, as well as on the contemporary redefinition of British urban policy and emergent forms of urban and regional governance. He was coauthor (with John Allen and Doreen Massey) of *Rethinking the Region* (1998) and joint editor (with John Clarke and Sharon Gewirtz) of *Comparing Welfare States* (2001) and (with Deborah Talbot) of *Security: Welfare Crime and Society* (2008). His book *Understanding Urban Policy: A Critical Approach* was published in 2007.

S. Harris Ali is associate professor at the Faculty of Environmental Studies at York University, Toronto. His research interests involve the study of environmental health issues and the sociology of disasters and risk from an interdisciplinary perspective. He has published on toxic contamination events and disease outbreaks in such journals as *Social Problems, Social Science and Medicine, Canadian Review of Sociology and Anthropology, Journal of Canadian Public Policy, Urban Studies,* and *Geography Compass.* He has recently coedited (with Roger Keil) a volume entitled *Networked Disease: Emerging Infections in the Global City.*

Roger Keil is director of the City Institute and professor of environmental studies at York University, Toronto. His research focuses on urban governance, global cities, infectious disease and cities, urban infrastructures, and urban political ecology. Among his recent publications are *The Global Cities Reader,* edited with Neil Brenner (2006); *Networked Disease: Emerging Infections in the Global City,* edited with S. Harris Ali (2008); *Changing Toronto: Governing the Neoliberal City,* with J. A. Boudreau and D. Young (2009); and *Leviathan Undone? The Political Economy of Scale,* edited with Rianne Mahon (2009).

Doreen Massey is professor of geography at the Open University. Her long-term research interests focus on various aspects of the theorization of "space" and "place." These interests range through critical engagements with globalization, regional uneven development, and the reconceptualization of place. Her work has been pivotal in conceptualizing what she has termed the "global sense of place," which has motivated much thinking on global relationality and cities. She is author of numerous books and articles, including *Space, Place, and Gender* (Minnesota, 1994), which includes her classic essay on the global sense of place; *For Space* (2005); and *World City* (2007). She is also cofounder and coeditor of *Soundings: A Journal of Politics and Culture* and a regular contributor to public debate, especially through BBC Radio.

Eugene McCann is associate professor in the Department of Geography at Simon Fraser University. He is an urban geographer with interests in urban politics, policymaking, and the relationships between urbanization and globalization. His contribution to this volume reflects his current primary research focus: a social, spatial, and political analysis of the development and deployment of harm reduction approaches to urban drug policy. This account conceptualizes urban harm reduction as a policy assemblage that is produced in and through activism and mobilities that stretch from the body, to the neighborhood and city, to the globe. He is coeditor of special editions of the *Journal of Urban Affairs* (2003, with Deborah Martin and Mark Purcell) and the *International Journal of Urban and Regional Research* (2006, with Kevin Ward) on the relationships between urban politics and space. His work has also appeared in the *Annals of the Association of American Geographers, Antipode, Environment & Planning A, Environment & Planning D: Society & Space, Geoforum, Professional Geographer, Social & Cultural Geography, Urban Geography,* and *Urban Studies.*

Donald McNeill is professor at the Urban Research Centre, University of Western Sydney. His research is concerned with the production of major commercial structures in urban space, such as airports, skyscrapers, and hotels, and the globalization of architectural practice. His contribution to this collection draws on research conducted as part of an Australian Research Council Discovery Grant, "The Production and Contestation of Airport Territory." He has written numerous research papers and several books, including *The Global Architect: Firms, Fame and Urban Form* (2008), *New Europe: Imagined Spaces* (2004), and *Urban Change and the European Left: Tales from the New Barcelona* (1999).

Jamie Peck is Canada Research Chair in Urban and Regional Political Economy and professor of geography at the University of British Columbia. He is a geographical political economist with research interests in economic regulation and governance, labor markets and employment/ welfare policy, and the politics of urban and regional development. His publications include *Work Place* (1996), *Workfare States* (2001), and *Constructions of Neo-Liberal Reason* (2010).

Jennifer Robinson is professor of urban geography at University College London and Honorary Professor in the African Centre for Cities at the University of Cape Town. Her recent book, *Ordinary Cities: Between Modernity and Development* (2006), offers a postcolonial critique of urban studies, explaining and contesting urban theory's neglect of cities of the global South. It argues the case for urban studies to draw on the diversity of urban experiences across the globe in developing more general accounts of cities. She is currently working on regrounding comparative methods to support a more properly international urban studies, and developing a comparative research project on the politics of city strategies.

Kevin Ward is professor of human geography in the School of Environment and Development at the University of Manchester. His research focuses on the geographies of urban policymaking, state reorganization, the politics of urban and regional development, and labor market restructuring. He is a coauthor of *Spaces of Work: Global Capitalism and the Geographies of Labour* (with Noel Castree, Neil Coe, and Michael Samers, 2004), *Urban Sociology, Capitalism, and Modernity* (with Mike Savage and Alan Warde, 2003), and *Managing Employment Change: The New Realities of Work* (with Huw Beynon, Damian Grimshaw, and Jill Rubery, 2002); coeditor of *Neoliberalization: States, Networks, Peoples* (with Kim

England, 2007) and *City of Revolution: Restructuring Manchester* (with Jamie Peck, 2002); and author of sixty-plus articles and book chapters that have appeared in venues, including the *Annals of the Association of American Geographers, Antipode, Environment & Planning A, Environment & Planning C, Geoforum, IJURR, Transactions of the Institute of British Geographers,* and *Urban Studies.*

INDEX

(continued from page ii)